WINGS OF TOMORROW

THE STORY OF

THE AUTOGIRO

Originally published in 1931

ISBN: 978-1-940001-44-9

The Aviation Collection
by
Sportsman's Vintage Press
2015

JUAN DE LA CIERVA

Photograph by Bachrach.

WINGS OF TOMORROW

THE STORY OF

THE AUTOGIRO

By JUAN DE LA CIERVA and DON ROSE

BREWER, WARREN & PUTNAM

CONTENTS

FLYING BY MACHINERY

EVER since man first made himself at home upon the earth he has watched with envy the flight of birds and wondered at their easy mastery of the air. Yet in spite of all his wit and accumulating knowledge, a thousand centuries were not enough to teach him their secret, though flying creatures have been his companions since the dawn of human history. But the bird's example and the increasing needs of men have kept alive an aspiration as ancient as the race itself—the desire to fly freely, speedily and surely in the open air, high above the crowded and tangled earth, straight to the desired destination, swift as the wind and skilfully as the swallow that darts and twists in the skies of evening.

This desire is written here and there in the legends of all peoples—the legends which are the true history of human hopes and ambitions. The winged horse, the flying carpet, the marvellous chariots that were driven by the will of their riders; the devices of magic which made nothing of time and distance. And here and there in later times were bold attempts to copy in more practical fashion the skill of the bird and the mode of its flights, sometimes with wings and sometimes with contrivances like giant kites or strange boats to be driven by oars through the ocean of air. The more daring among the inventive geniuses of other centuries could not resist the challenge of the bird. But neither could they discover its answer.

It is a very ancient quest that reached its goal in the early years of the Twentieth Century. Victory was so long delayed, indeed, that there were far more sceptics than believers when the tale of old Daedalus came true within the days of this generation. The general public of thirty years ago had less faith in the future of human flight than more primitive peoples, who cherished a dream without being much troubled by its difficulties of achievement.

And yet flight became a probability and even a

certainty more than a century ago, when man first learned to multiply his powers to an indefinite degree by means of machinery. The inventors of old did not lack skill or patience or an example to inspire their efforts to fly, but they did lack power. It is as true today as it was when Icarus fell into the sea that a man has not strength enough of his own to do what a bird does so easily. The muscular strength of a bird's wing is enormous in respect to its weight; if man had as much in his arms and shoulders and the bird's instinctive skill to use it in manipulating an extraordinarily complex pair of wings, he would need no elaborate aircraft to go about his affairs in the air as he does on land, while the planning and contriving brain that is the human advantage over the rest of creation would enable him to beat the bird at its own business. But for lack of such muscles he must use machinery, and when once he had discovered the possibilities of extended and multiplied power in fuels and mechanisms he had at hand the means for solving the problem of flight, even though measurable success waited a full century beyond the birth of the machine age.

But though countless generations of men had watched the flight of birds and wondered how they did it, it was not the bird that taught the world to fly. Aviation made practically no progress, indeed, until it gave up the inspiration of the bird and was thinking and working in terms of "flying machines." These were supposed to be possible long before they became practical. The notions of the romancers of the Nineteenth Century were borrowed very naturally from types of transportation that had reached some degree of perfection; they thought of boats of the air with great propellers or with paddles driven by engines far more powerful than were actually available. But they were thinking in terms of flying machines, not of birds. They were contemplating aircraft in which the flier was the directing and controlling genius, master of powerful wings driven by engines and mechanisms. The machine would fly; the man would manage it.

So science waited a long time for engines and mechanisms powerful enough in proportion to their weight for the purposes of flight. These came into existence with the development of the internal combustion engine. Meanwhile there had been many men of genius working with kites and gliders, determining that the swift motion of a broad surface through the air or the similar use of wind currents would sustain the weight of a man and a machine. Two elements in scientific research were moving mutually toward a practical flying machine—the perfection of light-weight power plants and the discovery of fundamental aerodynamic laws. They were combined at last by the Wright brothers in the first aircraft to leave the earth under its own power, fly through the air and land safely.

What the Wright brothers built in 1903 which had never been built before was a flying machine. It was also a kite and glider; it was in no respect a bird or anything like a bird. It flew by means of a motor and surfaces which were called wings, but which were actually totally unlike the wings of a bird. They were fixed planes, built as rigidly as possible within the existing range of knowledge. And though the pilots of aviation's early days were popularly called "birdmen," the craft in which they flew was called an aeroplane,—a fairly accurate term for a flying machine whose chief characteristic was its broad fixed surfaces or planes, propelled or pulled through the air by a gasoline motor.

The aeroplane, whose name was soon shortened to airplane, was successful because it was a flying machine and not merely a man-made imitation of the natural wonder of the bird in flight. But a curious conservatism seems to have surrounded its history during the past quarter century, unlike the energetic uncertainty which has inspired progress in most of the mechanical arts. The first flying machine was an airplane; it seemed for many years that all such craft must be airplanes, for there were few signs of promise and none of success in any other applications of mechanical principles to the problem of flights. There were some scattering efforts to build helicopters and ornithopters, but none that proved remotely successful. In the province of heavier-than-air flight the airplane seemed supreme and nearly all the world's engineering and aeronautical research was devoted to its perfection.

It was by a return to first principles that the Autogiro was discovered and developed. Juan de la Cierva, an ambitious young engineer, well started on an aeronautical career and looking ahead to the perfection of the plane, conceived the idea of looking in the other direction to the beginning of the science of flight. These beginnings had led by stages to the airplane of his time; he resolved to consider whether they might not have led to something else and something fundamentally better for the purpose of mechanical flight.

This was the real genesis of the Autogiro. Its inventor was persuaded by experience and convinced by his studies that it was wrong to assume that the only practical heavier-than-air craft was the airplane. He went back to the fundamental idea of the flying machine, recognizing that the conventional airplane is one of its types but not necessarily the only one. Not even, indeed, the best one, for the approximate perfection of the airplane had resulted in a flying machine of definitely limited performance, efficiency and dependability. It seemed to Juan de la Cierva, contemplating the history of aviation and a crisis in his own career, that aeronautical science had been prejudiced from the beginning in favor of the airplane and had more or less ignored the possibilities of other applications of the principles of flight, though experience had revealed a number of inherent deficiencies in the plane which persistently baffled the ingenuity of the inventors and designers.

The success of the Autogiro proved the soundness of his reasoning. It justified the opinions expressed by prominent pilots and engineers in its early days,—that this was the only new invention in aeronautics since the triumph of the Wrights. It was new from the beginning, not merely in some contributory detail of design or mechanical improvement. It was not a modified airplane; it was a new sort of flying machine. It has proved itself a better flying machine in most important respects than any of its competitors.

The Autogiro is not an airplane, though it performs the functions expected of an airplane, employs all its experience and has adopted most of its devices, instruments, materials and manufacturing methods. One airplane may be compared with another in respect to performance, efficiency or utility; the Autogiro must be compared with all airplanes to discover its essential characteristics

EVERY FIELD A LANDING FIELD

The Autogiro cruising over Pennsylvania, with a thousand landings in sight.

FROM GLIDER TO BOMBING PLANE

Above: Juan de la Cierva's first glider.
Below: Tri-motored bomber entered in Spanish Government contest.

and its proper place in the theory and practice of aeronautics. And by comparison it may be determined whether the airplane or the Autogiro offers the greatest promise of useful perfection in the future. It will be discovered, indeed, on the basis of present evidence and experience, that the Autogiro is the flying machine that answers the needs of the average air-minded man of today, as the airplane does not.

The distinction between the airplane and the Autogiro is fundamentally important, both to understanding of the newer type of flying machine and to appreciation of its possibilities. It is not so necessary to insist on the differences between the Autogiro and the helicopter, principally because the history of the helicopter is almost entirely a history of good intentions gone wrong. The obvious advantage of aircraft which would rise and descend vertically, hover under control in mid-air and at the same time be capable of full forward flight has not inspired much success in their development. A few helicopters have flown; none have threatened any serious competition either to the airplane or to the Autogiro.

The helicopter is a flying machine designed to rise by means of a vertical air screw or propeller. It is made to pull itself vertically into the sky, as the airplane's propeller pulls it forward in the line of flight. This means, of course, the application of mechanical power to a propeller or propellers in the horizontal plane. A simple difference between the helicopter and the Autogiro is that the revolving wing which looks so much like an oversized propeller above the Autogiro is not power-driven, except in the process of starting while the craft is still on the ground. It is in no sense a propeller. Therefore the Autogiro is in no sense a helicopter.

The Autogiro is, therefore, neither airplane nor helicopter. Nor is it a mechanical bird, such as satisfied the imaginations of the ancients. It is a practical and fundamentally perfected device for flying by means of machinery. It is a true flying machine. So, indeed, is the airplane, and so would the helicopter be if it would fly as well as either the airplane or the Autogiro.

It may be that there are other types of flying machine unknown to present knowledge and experience. There still remains the theoretical possibility of one in a machine which would imitate the motions of the bird's wing, but no such machine has ever been suggested which showed the slightest promise of successful and efficient flight. It is a matter of aeronautical science and experience, indeed, that it is time wasted to attempt to imitate by mechanical means the flight of birds, just as it would be absurd to pattern an automobile after a galloping horse. Machinery does not imitate nature; it utilizes the laws of nature but its means and methods are its own. And the future of flight is a future of flying machines, whatever their triumphant type may turn out to be.

So when Juan de la Cierva surveyed all the reasonable possibilities of mechanical flight, he was looking not for a mechanical bird but for a better flying machine than the airplane and found it finally in the Autogiro. He could do so only because of the comparative maturity of the machine age, which had already made possible the airplane and its amazing performances. Aeronautical science was sufficiently advanced at the time of his earliest speculations to have eliminated nearly all uncertainty in regard to adequate power, structural strength and the essential accessories to a successful flying machine. The airplane motor was approximately perfected, the descendant of the steam engine by way of the gas engine and the automobile motor. The variety of metals and materials demanded by a world filled with machinery gave his experiments an immense advantage over the courageous ventures of the Wright brothers and those who preceded them. The science of aerodynamics had survived a long period of trial and error and been reduced to formulæ, equations, wing curves and known laws of lift, drag and stress. Adequate instruments were available, practical devices for control and numerous safeguards against the host of hazards that beset the earlier designers, inventors and fliers.

What was lacking was the touch of inventive genius which would discover a better highway to the success that everyone was seeking. Everybody was seeking the same sort of success, for aviation has never been in doubt as to what is wanted and needed to make flying useful and universal. Ample speed for full flight, slow speed for landing, easy and effective control under all flying conditions, inherent stability and security, the ability to land and take off in limited areas,—these are aviation's ideals and the practical problems of aeronautical science. Some were solved in the perfection of the airplane; some were still far from solution after years of intensive and expensive effort. The airplane was perfected in a thousand respects, but the more it was developed the more clearly it revealed its limitations.

The outstanding inspiration of the Autogiro's story was the recognition that there might be more than one type of flying machine, soundly mechanical but basically independent of conventional theory and practice. Many men, of course, had thought of this possibility and are still considering it; one had wit and skill enough to discover it. And it was done not in a moment of inspired lunacy such as has been sometimes responsible for a great idea, but by a calm and scientific consideration of all possibilities, the elimination of those which would not work, and the final concentration upon a principle which could be made practical.

That is the story of the Autogiro. It begins with

an episode of failure—the crash of an ambitious airplane in sight of its designer, who had built into it his years of study and experiment, his youthful ambitions and his financial resources. It progresses though discouragements and disappointments and persistent problems, which yield at last to determination and conviction in the soundness of a scientific idea. It ends in the promise of a new lease of life for aviation, a great industry built around a new sort of flying machine, a new and universal utility of transportation and human convenience. It is the story of a machine and also of a man, Juan de la Cierva of Spain,—scientist, mathematician, inventor, pilot and an outstanding personality in the profession of his choice.

But at bottom it is a new romance of the modern world, the world in which machines are made to do men's bidding. It is the history of the flying machine of the future and the wings of tomorrow, which are already soaring in increasing numbers along the airways of today. It is the story of a piece of machinery, but of one that answers the ambitious quest of humanity through many centuries. It is a story of wings,—strange wings like those of a flying windmill—by which man may master the air and use it freely for profit and pleasure, for business and amusement, for varied needs and uses in today's swift civilization.

Don Rose

I

SCHOOLDAYS IN SPAIN

IT is no easy matter to trace an idea all the way back to its initial impulse. An ambition, an idea or an invention are likely to get their start in some inconsiderable incident or apparently trivial experience, or will originate in an interest that seemed at the time undeserving of serious attention. And I suppose that many familiar factors in our modern life might be traced to a schoolboy's mischievous experiments or to the undisciplined imagination of youth.

Certainly the story of the Autogiro is nearly as long as the life of its inventor. I was only fourteen when I was first taking myself very seriously as an aeronautical designer. That was in 1910, when I was still a schoolboy, one of a group of equally irresponsible youngsters interested in the fascinating possibilities of human flight. Probably two or three of us were ringleaders in our experiments; I believe I was considered the chief designer of the company. I recall, moreover, the impression that our parents were not particularly pleased with our choice of a means of self-expression.

Nowadays boys are likely to be fairly well-informed on aeronautical matters long before they are finished with their education. Their knowledge is sometimes, indeed, rather remarkable; they seem to take readily to the technical terms of the trade, they are often familiar with types of plane and engine, they make critical comparisons and estimates. Their heroes are likely to be the outstanding pilots of their time. Many of them amuse themselves by building model airplanes that are often remarkable in accuracy and workmanship. In nearly every country they compete among themselves in building and flying model aircraft. They accept the fact of aviation almost casually and have no such fears of flying as are likely to embarrass their elders. The parents of today, indeed, must sometimes feel themselves far behind the younger generation in understanding and acceptance of the new mode of transportation.

But that was not the case in 1910. That was years before the War, which forcibly impressed the fact of aviation on the world at large. It was only seven years after the first flying machine lifted itself into the air at Kitty Hawk. I was only eight years old when that momentous event occurred, for I was born at Murcia, Spain, on September 21, 1895. The Wright brothers brought their experiments to their first real success in December, 1903. It seems significant and a little startling to realize that the whole history of human flight in heavier-than-air machines falls within the compass of a young man's lifetime.

But my interest in aviation began, more or less, with my first real interest in anything. My first self-selected studies were devoted to the elementary science and art of human flight. While still a high school student in Murcia and Madrid I was learning all I could of the work of Langley, Maxim, Ader, Chanute and others who had done such important pioneer work with gliders. Particularly I was impressed with the theories of Otto Lilenthal, who was in some respects the real pioneer of practical aviation. When he was killed in 1896 he had made more than two thousand gliding flights and had determined the essential principles by which a powered machine might be made to fly.

I studied his theories and designs with patience and persistence. There was not much more, indeed, that I could study. I doubt that Spain owned many airplanes in 1914 and certainly none had been built in that country before that time. We rarely saw an airplane in flight, though we read avidly of what was going on elsewhere in the world. So we dug diligently into the library's resources and did our best with them. And with the supreme confidence and impudence of youth we were soon determined to put our knowledge to the test of practical experience, by building gliders of our own and flying them if we could.

We were not much encouraged in this by our elders. My father was busy with public affairs, a leader of the Conservative Party and twice a member of the Cabinet. After the disorders of early 1931, he was again appointed to office by King Alphonso, who has long been a friend of the family. In 1914, I doubt that Juan de la Cierva Senior approved of his son's curious interest in kites and gliders, or encouraged his ambitions to build and fly them. But boys find ways and means to satisfy their extraordinary impulses, and we spent much of our spare time and most of our spare cash in these primitive experiments in practical aeronautics.

There were three of us who took our aviation seriously. One of the triumvirate was the son of a builder and contractor, whose plant included a well-equipped carpentry shop. This was the factory and laboratory for most of our work, not only in building gliders but in the subsequent restoration of a smashed airplane which represented the second stage in our aeronautical progress. I suspect it supplied us, moreover, with a proportion of our materials. My friend's father was apparently an indulgent parent.

Here we built several gliders, more or less according to our studies of the experts, with modifications of our own devising. Most of them were really kites of various types, which we tried out on the hillsides around Madrid. We did more, I think, than imitate our models and test the results in flight; we were also trying in profoundly amateur fashion to figure out the aerodynamic laws involved in the construction of kites of various shape and structure. We worried a lot about weight and we considered with the seriousness of extreme youth the obvious problems of building stamina into our flimsy fabrications. And in the natural course of events we came at last to the point where nothing would entirely satisfy us but a real flight in a passenger kite or glider.

Fortunately for our safety and the continuance of our experiments, we had no high cliffs at hand for our take-offs. The best we could do was to slide off from a fifteen-foot bluff, below which lay a gently sloping hill. I think I was the first pilot of our primitive aircraft. It had some sort of controls, crude ailerons and rudder, and the motive power consisted of a company of small boys from the neighborhood at the end of a few hundred feet of rope. We must have learned something by our studies in theoretical aeronautics, for the thing actually flew. With me in the pilot's seat it never rose above four or five feet, but there was no doubt that it was a flying machine, able to carry its own weight and a fairly substantial passenger. And that first flight, I assure you, was the thrill that comes but once in a lifetime. I remember it still, as I remember the thrill of ten years later, when the first successful Autogiro lifted from the ground and flew according to my expectations.

One glider flight led to another and eventually to an accident. It seemed to my younger brother, who had done his share of pulling at the rope in order that I might fly, that he was entitled to take a turn at the controls. The three proprietors of the glider discussed the suggestion at length, and at last agreed to give him a ride. Carefully I explained the controls and put him through ground tests in their management. I was aware of my responsibilities and probably self-conscious of my superior aeronautical knowledge. But I overlooked entirely an important factor in the situation.

My brother was both younger and lighter than myself. This upset all my aerodynamic calculations and resulted in a new set of performance characteristics for the glider. Under the impetus of willing hands on the rope, it shot suddenly fifty feet in the air, brother and all, which was both unprecedented and unexpected. It was so startling, in fact, that the engine stopped, which is to say that the boys on the end of the rope forgot their vital share in the demonstration and stopped to stare and applaud. Thereupon the glider ceased at once to glide, and came down in a spin as suddenly as it had gone up.

My brother went to bed for a few weeks to recover from the effects of his first flight. He was superficially damaged but not badly hurt, and we managed to keep the secret of his accident, reporting to the family that he made a bad landing from a bicycle. Not until fifteen years later was the real story revealed, and even at that distance my father's consternation was considerable.

Another minor accident put an end to our glider experiments. In 1912 I had built a free glider, a biplane of the Chanute type, which we tried out from a convenient hilltop. While another of the trio, Pablo Diaz, was trying to pull it into the air, a gust of wind lifted both the glider and the boy to a considerable height and dropped them again as precipitately. On the way down he let go of the glider. He landed sitting, with a formidable and painful thump, and the glider blew over the hills and far away. And that was the last of our gliders. Thereafter I continued my studies and experiments with literally thousands of paper planes, reading meanwhile every word on aeronautics that I could lay my hands on.

My interest in aeronautics, indeed, was now settled and seriously intentioned. Most of the other boys were only interested bystanders to our ventures and adventures and their attention turned elsewhere when we gave up our public demonstrations. The three of us continued to consider ourselves as prospective designers, engineers and pilots. Considering the instable condition of aeronautical science at the time and our meagre sources of information and contact with what was going on, I think we knew quite a lot about flying and flying machines. Our knowledge was almost entirely theoretical, but it was fairly accurate and well founded on good authorities.

My own interests were primarily mechanical and mathematical. Aside from my school studies, I found time and patience to pick up a certain amount of mathematics, physics and engineering science. I did not know it at the time, but it seems now that I was to some extent determining my destiny in pursuing these hobbies, for that is what they seemed to be to a fourteen-year-old schoolboy. I had no thought of the Autogiro, of course, but very likely I would never have arrived at it if it had not been for this early interest in practical and theoretical aeronautics. I was thinking then, in the optimistic fashion of a youngster, of a career in aviation. Very likely this was no secret to my parents, for later they allowed me the advantages of a full course in technical engineering and were sympathetic in very practical fashion to my ambitious experiments.

But before I reached the stage of a settled attention to technical studies, my two companions and I had another adventure which quite overshadowed our glider experiments. We built an airplane, or at least rebuilt it from a heap of wreckage into something that would fly. And it was, so far as I know, the first airplane ever assembled in my native land.

II

SPAIN'S FIRST AIRPLANE

THERE was living in Madrid at this time a French civilian pilot named Jean Mauvais. He made his living by selling automobile and bicycle accessories to the general public and aviation supplies to the Spanish government, but he was also the proud proprietor of an early model Sommer biplane, with which he gave occasional exhibition flights at Madrid and elsewhere.

He was a genial gentleman. I remember that he used to travel to the Cuatro Vientos Airport, our favorite playground in free hours, in an ancient Dion Bouton single-cylinder automobile. The field at that time had a few wooden hangars and some canvas sheds, in which were stored a few biplanes of the pusher type. There was also a very tall

TWO PIONEER PLANES

Above: The "Red Crab," Spain's first airplane.
Below: Monoplane built by Juan de la Cierva.

TWO "UGLY DUCKLINGS"

Above: First Autogiro, with double rotor, which failed to fly.
Below: Second unsuccessful Autogiro with three rigid rotor blades.

steel structure, fabric covered, intended to house an early type of dirigible built in France for the Spanish government. When the airship was not in active use, this building served as storehouse for wrecked aircraft and a variety of aeronautical odds and ends.

The government fliers, having very often nothing better to do, inflicted all kinds of practical jokes on M. Mauvais, and were particularly fond of hiding his tiny two-seater car just when he was ready to ride home in it. An airplane packing case was big enough for the purpose, but Mauvais soon became familiar with such hiding places and could commonly find his car without much trouble. On one occasion, however, he hunted it for hours. We helped him find it at last, hung high in the ceiling of the dirigible hangar by a rope.

We became, indeed, rather friendly with the French pilot, and it was he who unintentionally gave us our biggest opportunity to date in the building of aircraft. One day he gave a demonstration flight at a race course near Madrid. As he came in to land, an enthusiastic crowd swarmed on the field and rushed toward the descending airplane—a not uncommon occurrence in the days when the public had not learned to respect the dangerous possibilities of a whirling propeller or to appreciate the difficulties of landing a plane. In the effort to avoid hitting someone, Mauvais wrecked his ship, though not without killing several spectators. The pilot was practically unhurt, but he said he was through with flying. He dumped the wreckage in the rear of his little workshop. Not much was left except the Gnome rotary engine and the landing wheels; the rest was a tangle of splintered wood and shredded fabric.

We entered into negotiations with M. Mauvais. It was our modest proposal that we should take over the *débris* and build a new airplane with it. We would return the airplane to the pilot on condition he would test it for us when it was reconstructed. We promised, for good measure, to incorporate into the rebuilt machine a number of our own ideas and improvements. Probably the pilot thought this an excellent joke; very likely he supposed the bargain was safeguarded by the likelihood that we would never complete our part of the contract. In any case, he agreed.

Our joint capital was about sixty dollars. We had also the resources of the carpentry shop owned by my friend's father. In addition there was available any amount of youthful ambition and self-confidence. So we began to rebuild the airplane. We could buy no new parts, owing to our financial limitations and the scepticism of our parents as to the importance of our undertaking. So we made them as best we could. We whittled out struts and spars, begged help and tools from the woodworking department of the contracting business and sometimes hired a carpenter by the hour to assist us. We got along nicely enough, indeed, until there appeared the problem of replacing the smashed propeller.

I knew that propellers should be built of laminated wood—this was before the days of steel propellers—but we had no proper wood and we couldn't have laminated it if we had. To buy a new propeller seemed out of the question. But a propeller we must have, or else our airplane would never fly.

I pondered the problem at length and was rewarded at last with an inspiration. It seemed to me that seasoned wood was a prerequisite to a successful propeller. We looked around for seasoned wood of a suitable consistency. I forget exactly where we found it, but it was in the bar-room of a little inn, where the counter had for years been soaking up the alcoholic remnants of spilled beverages. I reasoned with my companions that the counter must be thoroughly seasoned. They could find no flaw in my argument, so we purchased the bar-room counter and carried it off to our factory.

Out of it we chopped and chiseled our propeller. I thought I knew enough of theory and design to give it a respectable curve and balance, and it actually worked well enough in flight. No doubt it wobbled and vibrated a little, but a few vibrations more or less made little difference to the performance of our craft. The propeller lasted, in fact, as long as the airplane, and I think it should go on record that the first airplane built in sunny Spain flew successfully on the wine-soaked table-top of an old-fashioned bar-room.

Our other main problem was to find proper fabric to cover the wings and control surfaces. Probably it could not have been purchased in Spain nineteen years ago, and in any case we had hardly any money. So we bought the cheapest canvas, stretched it as well as we could by hand, and doped it liberally with glue. Then we colored the wings and fuselage with aniline dye to a satisfying scarlet, christened our craft "The Red Crab," and were ready for the test flight.

I admire the courage of M. Mauvais, who found out that "The Red Crab" would fly. But he had warmed to our work as it progressed, and had contributed valuable advice and assistance. He may not have shared our supreme confidence in our craft, but he was willing to take it for a ride. We had rebuilt it as a pusher biplane, but we had enlarged the cockpit to make room for a passenger and changed the design so that the ship could carry one. It did so successfully and for some time. I flew in it many times. Sometimes I would sit behind the French flyer and reach around him to hold the wheel, but I made no mistake of thinking myself a pilot.

Thereafter "The Red Crab" made many flights around Madrid and carried many passengers.

M. Mauvais' interest in practical aviation, indeed, seemed completely recovered, and he did quite a brisk business. It looked as though "The Red Crab" would last a long time. But the glue was literally its undoing. Whenever it rained, the ship became sticky all over, and after a while it began to come apart. Nobody minded much that its wings flapped a little in flight, but it was serious when they began to fall to pieces. But "The Red Crab" never crashed; it simply disintegrated.

Our families seemed impressed by our successful venture and liberalized our allowances for aeronautical experiment. In 1913, therefore, we began the building of a fast monoplane with a new Le Rhone motor. In our desire to secure speed we gave the ship so little wing surface that the motor would not take it off the ground, and it cost quite a lot to redesign and rebuild the wings. When it finally flew, it was too fast for easy handling. It was tricky to get off the ground and trickier to get down again, but in the air it flew like a racing plane.

But it had many minor accidents and our parents tired at last of paying for the rebuilding of a wing tip or a new propeller. Every airplane costs money to maintain, and ours was no exception. Finally, when a landing gear collapsed in taxiing, this served as an excuse to shut down altogether on our financial support. The ship was sold and our triple alliance broke up.

My two companions continued their interest in aviation and engineering. Jose Barcale became a civil engineer and Pablo Diaz continued his association with me in subsequent experiments. For many years, indeed, he was foreman of my experimental shop.

In the year 1912 I was enrolled as a student at the Civil Engineering School in Madrid. There were at the time no aeronautical schools in Spain, nor any regular college courses available in the subject. When I graduated, indeed, it was with the title of Ingeniero de Caminos, Canales y Puertos, which is considered the highest degree in engineering in Spain. It was not until 1930 that a special University for Aeronautics was created in Spain, at which time I was granted the title of Aeronautical Engineer *honoris causa*—that is, without examination. This distinction I have the honour to share with only one other man, Torres Ginevedo, the famous engineer.

But in 1912 my ambitions were fairly definitely determined, though my studies were not specifically arranged to suit them. Subsequently I was grateful for a general education in engineering and for the varied studies of those six years in the Technical College. But from the beginning, I believe, I was more interested in aeronautics than in "railroads, canals and ports." And I took advantage of every opportunity to experiment further in the practical problems of my favourite science.

Then came the War, from which Spain seemed so far away but which stirred us air-minded youngsters with tantalizing hints of aeronautical progress everywhere else in the world. The literature of aviation increased and its science began to be incorporated in text-books and treatises. I read all I could get, still a student of civil engineering but more determined than ever on a career in aeronautical designing. There were no more practical experiments, but I worked out many theories and drew many designs, testing some of them with models and comparing them with what I could learn of progress elsewhere. Some of my ideas were quite unsound; others were substantiated later by accepted developments in airplane design. I was particularly fascinated at the time by the study of wing curves and airfoils, and mastered as much as I could of their complicated theory and formulæ.

Meanwhile my natural bent for mathematics found plenty to do in the advanced courses which completed my education. This turned out to be important to me later, when I was obliged to work out the theory of flight by autogyration before I could put it to any sort of a practical test. It would have been virtually impossible, indeed, to have achieved any success with the idea of the Autogiro unless I had been able to calculate its basic design by mathematics before I began to build it. Success by mere experiment would have been as unlikely as the successful construction of a cantilever bridge without any previous engineering experience. More so, indeed, for the problem was far more complex and the risks of error far greater. To paraphrase a familiar American phrase, I had to be sure I was right before I dared to go ahead. Without a little certainty of science on my side, I might have spent my energies in a dozen different directions, all of them wrong, and never found the secret of the Autogiro.

Six years of schooling and my studies on the side first inspired me, however, to one more venture in conventional airplane design and construction. It happened in 1918, while the War was coming to its climax and end. While the War was on, no nation was distributing photographs, let alone technical data, of its military aircraft. Spain had done little or nothing to keep pace with aeronautical progress elsewhere, but finally resolved to encourage native talent with an open competition for the construction of three types of military aircraft—pursuit, reconnaissance and bombing plane. The competition was under the auspices of the army; the prize was ten thousand dollars and the privilege of supplying further ships to the government.

With the aid and encouragement of a little group of enthusiasts, particularly of Conde de los Moriles, I decided to enter the contest. There were a number of entries in the classes for light and fast craft —the pursuit ships and reconnaissance planes.

But there seemed to be no competition whatever in the bomber class.

I resolved, therefore, to design and build a big bombing plane.

III

THE CRACK-UP OF A CAREER

SIX years of serious studies and more of keen interest in aeronautics had brought me to the year 1918, nearly ready for graduation from college and anxious to get started on a practical career. The competition proposed by the Army Department of the Spanish government was as good an opportunity as an ambitious young designer could ask. Success would mean quick recognition and an immediate chance to put my training and experience to work in what promised to be a profitable business. It seemed certain that the designing and building of aircraft for military purposes would be pushed rapidly ahead as a result of the War's lessons, whatever might come of their commercial possibilities.

The prizes offered were not of generous proportions, but the chance to contract for further deliveries of successful types was worth working for, though there was little prospect of any orders for bombing planes. But, above all, I was anxious to make a real beginning at an aeronautical career. So were others among my fellows, some of whom joined me in my venture and contributed financial assistance. But it was agreed with my backers that I should be the designer of our entry in the competition.

We built the bomber at the cost of about $32,000—a considerable sum now and more in 1919. It flew only a short time and then crashed beyond repair or redemption. Hardly a successful or profitable venture, but nevertheless the most fortunate accident of my aeronautical career.

It was a big biplane that we built, equipped with three tractor Huispanio Suiza 180 horse-power motors. It had a wing span of eighty feet and was supposed to carry the equivalent of fourteen passengers. It compared, therefore, with the larger transport planes that are in general use today and with the bombing craft in modern air fleets. It was so large, indeed, that we were obliged to tear down the walls of our workshop in order to get it out for its test flights. There was nothing modest about either our ambitions or our confidence in choosing this class of plane for our entry.

It was, I believe, the second tri-motor plane to be built in the world. An Italian Caproni tri-motor preceded it, a large craft with two tractor motors and one pusher which attracted a great deal of attention by its size and power. But our ship embodied many details of advanced design, particularly in the use of a thick wing and in the economical distribution of bracing struts between the wings which improved both its speed and appearance.

We chose as pilot one Captain Rios, an experienced army flier. He had fought against the Riffs in Morocco and was wounded there by a sharpshooter while flying low over the enemy. He was a thoroughly capable man, though most of his experience was with fast fighting planes. He had never flown anything as large as our biplane, and it was natural that at first he should handle it a little nervously and apprehensively.

But very soon he discovered that it was exceptionally manageable and responsive, and his caution changed to the other extreme. He began to take it round the test course as though it were a pursuit plane, much too close to the ground to provide a proper margin of safety and taking the turns with more spirit than judgment. As we watched, our big ship suddenly faltered and crashed to the ground. When we dug Captain Rios from the wreckage he was unhurt save for a few bruises and scratches, but the plane was hopelessly smashed.

Neither the pilot nor the spectators could make any criticism of the plane's performance nor detect any evidence that it had failed for structural or mechanical reasons. So far as we could determine, Captain Rios had become over-confident, misjudged by a little his minimum flying speed and got himself into trouble too close to the ground to get out again. Had the same thing happened while he was flying high, he could probably have recovered control and brought the ship safely down. Only the lack of a few hundred feet of open air had brought my most ambitious effort in aeronautical design to so convincing a conclusion. And I admit that at the moment it looked like the end of my career in aviation. Possibly it meant that I should content myself with civil engineering or find myself a place in my father's business.

But what struck me most forcibly at the time was the fact that a big and expensive airplane should come to such a bad end simply because it could not fly slowly in safety while close to the ground. There seemed to be a serious discrepancy or disproportion between the cause and the effect. A little mistake had done such a great deal of damage, and the mistake was one which could only be made in an airplane. In every other vehicle, safety seems to be implied in slow speed and in keeping close to a stopping place. But in the case of the airplane, the exact opposite was true. The pilot knew it, of course, and should have remembered it, particularly in handling an unfamiliar craft. But because he was guided by impulse or made an error in judgment, my airplane was eliminated from the competition and my friends and I were left with nothing but a lot of expensive experience to show for our studies and ambitions.

Fortunately for my own future, I managed to find a moral in my lesson. I found it again in a

TWO EARLY TYPES OF AUTOGIRO

Above: The third full-sized model with five rigid blades.
Below: The first successful Autogiro with articulated blades.

EVOLUTION OF THE ROTOR HEAD

Rotor head and mechanical starter of the P.C.A.-2 Autogiro.
Inset: The first articulated rotor head.
(The endpapers of this volume show the rotor head in detail.)

little incident which happened later at one of Spain's military airports. A certain general had screwed his courage to the sticking point and resolved to take his first ride in an airplane. He was accustomed, of course, to taking all precautions appropriate to a valuable officer; he was also accustomed to giving orders and having them obeyed.

All arrangements having been made for the flight, with a respectful audience at hand to watch the general get acquainted with the possibilities of aviation, he called the pilot to stand before him. The pilot saluted and stood at attention.

"Is everything prepared?" asked the general. Everything, it seems, was prepared.

"You are feeling fit?" asked the general. "And your airplane—it is in good condition and safe?"

The pilot expressed his confidence in his own good health and the soundness and stability of his craft.

"Then we shall go," said the gallant general. "But I command you to fly slowly and to keep near to the ground."

It is unnecessary to put on record the embarrassment of the pilot as he tried to explain to his superior officer that it doesn't do to fly airplanes slowly and near the ground. He disobeyed entirely, of course, the general's instructions. Probably he flew higher than usual and faster than necessary in order to safeguard his precious cargo.

But though the general was wrong, he was right in so far as his comment was a criticism of the airplane. All his experience was to the effect that it is safer to go slow and advisable to remain in ready reach of a stopping place. That is, indeed, nothing but common sense. Unfortunately it cannot be applied to airplanes, for they cannot go slow and they are most likely to get into serious trouble while they are within a few hundred feet of the ground. So far as could be determined, my tri-motor was destroyed because its pilot was more or less following the general's instructions, though these were not actually given for his benefit. He was apparently flying too slowly to maintain altitude and control, and too near the ground to escape the consequences.

This was a first lesson to be derived from the crack-up of my youthful career. Another was the high cost of error in developing and operating an airplane. This is a matter that has been too much minimized in estimating the progress of aviation. But I think it fair to say that my own experience was quite typical and in some respects a parallel to the whole history of the art and science of flight. The only really successful plane I built was "The Red Crab." My monoplane was experimental, expensive and eventually was abandoned as an unprofitable venture. My big biplane was sound in structure and design, but it succumbed to a comparatively trifling accident. Only "The Red Crab" flew until it fell apart from old age and weariness. And if there was ever a primitive piece of aircraft, it was "The Red Crab."

So I wondered then, as I wonder now, whether we have any right to claim so much for the progress of aviation if it has failed to meet its most apparent challenge. Experience showed that I was no nearer to absolute safety when I built the biplane than when we reassembled "The Red Crab." The bomber was immeasurably better built, sounder in design in a thousand details, and equipped with many ingenious instruments and devices. But it was ruined in a few seconds by exactly the same sort of accident as might have happened to "The Red Crab" or to any other airplane built before that time or since. In this respect one plane was as primitive as the other.

It is this fact that has made aviation pay so dearly for progress. It has cost the lives of many brave and useful men and an immense sum of money spent and gone in crashed airplanes. All progress has its price, of course, but rarely is the price so high. Every other development in transportation has paid early attention to safety; subsequent progress has depended on it and counted on it. Whenever assurance of security has been lacking, other matters have been considered comparatively unimportant until it was attained in a reasonable degree. But because the limitations of the best airplane of today are in essentials the same as those of twenty years ago, the business of aviation has borne an extraordinary burden of waste and loss, sometimes to such an extent as to shake seriously the public's confidence in its future.

Certainly my own confidence in the future of aviation and my career in aeronautics was badly shaken by what happened so quickly and conclusively to the big biplane. Many others may have felt the same when they have seen some slight accident or blunder end disastrously for plane or pilot or both. But in my own case the opinion was necessarily personal and the problem immediate. Should I go on with my studies, experiments and ambitions? Should I be content to do my best under the limitations apparently imposed on the science and practice of aeronautics by the very nature of the flying machine? Or should I seek other employment for my technical training and experience?

I chose to go on, of course, in the career of my choice, though for some years I acted as manager of my father's agricultural and manufacturing interests. I even went to school again in later years, for I had never learned to fly and could not do so even when I was building the early Autogiros, so I trained some years later in England and secured there my pilot's credentials.

In subsequent years, even while conducting my isolated experiments in Spain, I kept in close touch with European progress in aviation and in the science of aeronautics. I have been honoured with the friendship of many fine pilots and have profited by association with the outstanding designers of my profession. And I am grateful to be able to acknowledge the many courtesies of my associates in aeronautics and the keen interest shown in my experiments by many distinguished societies devoted to the progress of aviation. They have been equally cordial to the Autogiro since it was first offered in public demonstration.

But these developments were not even imaginable as I surveyed the wreckage of my biplane bomber and wondered what to do next.

Note: Juan de la Cierva has received many honours from the leading aeronautical associations of the world. In Spain he is a permanent consulting member of the Consejo Superior Aeronautica, and a member of the Junta Superior de Estudios y Pensiones para Extranjero, an institution for aeronautical research and for awarding scholarships for students to visit foreign countries. He is a member of the Association of Spanish Civil Engineers and of the Association of Aeronautical Engineers in Spain.

He is also an Associate Fellow of the Royal Aeronautical Society, Honorary Member of the A. I. D. A. in Italy, a member of the Société Française de Locomotion Aerienne, and a member of the British, German, French, Spanish and Belgian Aero Clubs. He is a Chevalier of the Legion of Honour in France and of the Order of Leopold in Belgium, and he holds the Cross of Alfonso XII in Spain.

IV

A PROBLEM IN THEORETICAL AERONAUTICS

THE career of an aeronautical designer does not necessarily come to an abrupt end when his airplane is wrecked, even though the plane be smashed beyond repair. The designers and engineers who have served the science of aviation through its first quarter century have been compelled, indeed, to endure many such disappointments and discouragements. If crashes and casualties had shaken too seriously their faith in the future of flight and the soundness of their theories, aviation would have moved much more slowly to its present achievement. It might, indeed, have been abandoned as a highway too hazardous for profitable progress.

But aviation has advanced by a process of trial

and error, by pioneer courage and determination, and by paying a price for every step in the right direction. Very many planes have crashed, very many bold ventures in design have ended in failure. In a wider sense the science of aeronautics and the industry of aviation have survived similar experiences. Big ambitions have come to a bad end; bold plans have gone awry. Even success has become so quickly inadequate to the demands of aeronautical ambition that it has been discarded as relative failure, and new goals have been set and attained. Therefore the industry has been required to maintain its progress against a pace of obsolescence scarcely matched in any other field of mechanical progress. Planes go quickly out of date; aviation has adjusted its program again and again to keep ahead of the realized ambitious of its designers and engineers. It has done so, because faith in aviation's future has not failed in the presence of failure nor been discouraged by partial success.

Contemplating the wreckage of my most ambitious effort to date in design and construction, I had plenty of precedent to inspire me to go ahead in a career of conventional aeronautics. The crash was no fault of the plane or its designer. Properly flown, it might have been a successful contribution to aeronautical history. It would probably have won the prize, for it had no competition in its class. Probably it was a first-class airplane for its time, and yet it had flown only a few minutes before it was destroyed.

A slight error in judgment on the part of its pilot had destroyed in a moment the product of months of work and years of study. Very likely it was this fact that impressed me most at the time of the crash. I knew that nothing was seriously wrong with my design or the plane itself. It flew successfully at the first trial, though the inexperience of the pilot with planes of such size, weight and power led him to handle it nervously. As he became used to its controls he went to the other extreme. Having flown it successfully at a fair height, he began to turn it around the test course too near the ground for safety, considering that this was a new and untried plane and that his experience had been entirely with single-engine aircraft. He made a turn at too slow speed, lost control and crashed into the ground. Fortunately he was not seriously hurt, but the plane was finished.

It was the pilot's fault. And yet on second thought it seemed to me that it was really unfair to blame the pilot entirely because he had failed by a fraction to estimate a safe flying speed in an airplane which he knew very little about. It was an error in judgment that brought him down, not negligence or incompetence. Certainly he did not intend to take chances with my airplane and his own life. He made a mistake, as everybody makes mistakes. But because he was flying an airplane, the mistake was a serious one for me and might have been very serious for him. If he had made a similar mistake in driving a car or steering a ship or running a railroad train it would probably have made very little difference. But because he was piloting an airplane, his mistake cost me a considerable sum of money and endangered his own life. Incidentally, it threatened to put an end to my aeronautical career or to make it difficult to go ahead in the face of such discouragement.

But plenty of designers have seen their airplanes crash in test flights and have used failure as a stepping stone to success. I should probably have followed their example if the crash had been caused by faulty design or some miscalculation in construction. I should have found further financing, re-designed my plane and tried again. That sort of persistence has been the secret of success for many designers and makers of airplanes. But in this case the plane was apparently all right. The pilot, moreover, was reasonably competent. It seemed to me, as I surveyed the wreckage of my ambitious venture, that I should seek further for an explanation of the catastrophe. A good airplane in the hands of a good pilot had no business to be turned so suddenly and conclusively into a useless tangle of fabric and machinery just because the pilot made a miscalculation of speed and distance.

That, I believe, was the germ and genesis of the Autogiro and it is still the fundamental argument in its favour. My biggest airplane came to a bad end because the human equation in its operation outweighed all my aeronautical skill and all the pilot's experience. That, in a word, is what is the matter with airplanes, as may be demonstrated with facts and figures. Under certain conditions they must be flown with something like perfection or they will not fly with safety. And in a word, that is why the Autogiro is their superior for the purposes to which it is specially suited. Under all ordinary conditions it can be flown safely with only average skill and it can survive an ordinary error in judgment without harm to itself or its pilot.

But I was a long way from the Autogiro when I chose this lesson from my principal experience in conventional aeronautics. All that seemed sure at the moment was that something was seriously wrong with the basic idea of the airplane or that a highly important problem connected with it was unsolved. I devoted my attention, therefore, to isolating this problem from among the many matters on which aeronautical science was still feeling its way. I endeavoured to give it its proper importance, both to the theory and to the practice of aviation. I had something like a test case before me in the brief history of my own plane and its abrupt ending. I had a personal reason for thinking the problem through, for I must determine whether to go on with my studies and experiments or look elsewhere for a career that suited my tastes and abilities.

In this mood it was not difficult to decide what was the matter. If my plane had been able to fly slowly under control of the pilot it would not have crashed in turning a corner. If it had been possible to land it at immediate notice and in close quarters when the pilot found himself in difficulties, there would have been no serious accident. That was all that was the matter, but it was enough to wreck a perfectly good airplane and risk the life of the man who was flying it. And it was also perfectly obvious that these inabilities were not confined to my own airplane. No plane that had been built to that time had been able to fly slowly in safety near the ground or land under control in a small space at a reasonably slow speed. Nor can any plane built since that time do either of these things. The limitations involved seemed then, as they seem now, inherent in the airplane itself. And they are limitations that have cost many lives and led to the destruction of many airplanes.

That was my first conclusion; the second was

that something should be done about it. Something fundamental, indeed, and not merely a modification or improvement in plane performance, for this would only reduce a danger without eliminating it. If sixty miles an hour is too fast for safety in landing a piece of heavy machinery, forty miles an hour is essentially no better, except that it reduces a little the pilot's problem. It is still a dangerous pace at which to touch the earth and try to stop. The only prospect of assured safety was in a landing speed which could be compared with slow speed in surface vehicles and which would lead as quickly to a complete stop under full control.

But that was by no means the entire problem. The limitation of the airplane is not merely that it cannot land slowly; it is a more serious matter that it must fly fast in order to stay in the air under the pilot's control. The minimum safe flying speed is the most unyielding of all the airplane's limitations. It is actually as unyielding as the law of gravity and is essentially a manifestation of that law in operation. At a certain minimum speed an airplane no longer flies; it begins to fall. That minimum speed is about fifty miles an hour in the most favourable types of plane; in others it may be ninety miles an hour or more. In some craft that were built for the seaplane races it was as much as one hundred miles an hour. Below that pace they are no longer flying; they are also falling.

To complicate the matter still further there is the fact that at less than flying speed an airplane's controls become comparatively ineffective and may fail altogether. That, of course, happens in a stall or spin. And that was why my big airplane crashed to the ground and was destroyed. It was flying too slowly to answer to the pilot's will and it was also falling, possibly at a flying speed of fifty miles an hour or more. But though this speed was too slow for safe flight, it was much too fast for a safe landing under the emergency conditions which confronted the pilot, even if he had been able to make the plane answer to rudder and aileron.

As I have already said, this was not a particular weakness of my own airplane. Every plane that was ever built acknowledged the same limitations, and every designer does his work with them in mind. The Wright brothers knew of them and so does the pilot of the best airplane that flies today. Certainly I did not discover them when I studied the results of my own experiments in airplane design. But I did try to estimate their importance and I did ask if they were inevitable and inescapable.

Apparently they are, so far as the airplane is concerned. Nothing has been done or discovered in twenty-five years to get very far away from them. Airplanes dare not fly slowly and they cannot land slowly. Slotted wings and other devices have modified to some degree the danger of losing control at minimum speed, but they have scarcely reduced that speed. Brakes have been applied to wheels to shorten the landing run, but the plane must still touch the earth at about fifty miles an hour or more, unless helped by a head wind. The limitations are still there, in spite of all the development and refinement of the plane.

Having identified the limitations I began to contemplate the possibility of avoiding and escaping them, as an alternative to whittling away at them in the hope of reducing them eventually to a much safer minimum. I soon found that I was facing a problem in theoretical aeronautics, and that in studying it I must not be prejudiced in favour of the plane or of anything else. I was compelled, indeed, to consider the theory of flight from its beginnings, not merely from the point which I had reached when my airplane came down. I found I must go all the way back to the essential theory of flight and discover whether it could be applied entirely differently to the development of a successful flying machine. I must consider everything, hoping to discover something that would work. I must particularly avoid the assumption that because the first craft to fly under power was a plane there could be no successful heavier-than-air craft except the plane.

The essential theory of flight can be reduced to a comparatively simple statement, though it becomes a highly complicated affair as it is presented in figures and formulæ. It applies to every flying machine; that is, to every type of aircraft except the balloon and dirigible. In a limited sense it applies even to these lighter-than-air craft.

The air is a fluid, as water is, but of a much

thinner consistency. As a fluid it will support nothing that has weight, as we measure weight at the surface of the earth and in the sense that a solid supports weight. Neither will water, indeed, and most solids will sink in it, even though they are only partly submerged, since the water is displaced to some degree according to their weight. We note, therefore, a difference between the vehicle that moves on the land and the one that goes by sea. The cart, locomotive, automobile or other land vehicle rests on the earth or rolls on it. The ship floats; it does not rest on the surface of the water.

When we say that a ship floats we mean that its weight displaces an equal weight of water until the vessel weighs nothing. Enough of its bulk remains above the surface to make it a useful means of transportation, because a cubic foot of water weighs considerably more than the average weight per cubic foot of the entire ship. The same is true of the balloon and dirigible except that their weight is almost exactly balanced by the weight of air displaced by the gas bags. In comparing boat and airship it is well to remember that the airship floats in the air like a submarine submerged in the sea, not on the surface like an ocean liner.

The airplane does not rest on the air as an automobile rests on the road, nor does it float there like a ship on the sea. How, then, does it stay up?

For the sake of simplification we may say that there are three factors or forces involved. The first and most important is displacement. An airplane, of course, displaces a certain volume of air by its bulk, but the amount is negligible compared to its own weight and this factor can be ignored. And yet displacement does much to keep the plane in the air. It is not, however, a static displacement; it is the displacement of an immense volume of air by a rapidly moving plane. Flying, for example, at one hundred and twenty miles an hour, an airplane wing passes through 176 feet of air in a second, the actual volume displaced depending on the spread and curve of the wing. This displacement results in "lift," sometimes described rather crudely as the creation of a partial vacuum above the wing by its swift passage through the air.

The second factor depends on the fact that fluids become resistant like solids when they are moved at sufficient speed. A jet of water under high pressure, for instance, can be hammered with a steel bar and the bar will bounce off. The same principle applies to the air, as may be realized from the fact that a tornado can smash a building flat as if with a giant sledge and can drive slender straws deep into a piece of wood. The air passing beneath the wing of a plane in full flight cannot be considered as a solid, but it is at least resilient and resistant. It is estimated that it is as much so as the water to a swimmer moving at three miles an hour. And on this cushion of moving air—moving, that is, in relation to the flying wing—an airplane rests a proportion of its weight. About two-thirds of the lift is derived from the displacement of the air and operates from above the wing; one-third is support by air pressure from below.

Both these elements in an airplane's "lift" depend on its speed. So does another that is equally important. The air, we have said, is a fluid. As such, it moves rather readily out of the way of a solid, flows over it and quickly equalizes the pressures around it. It does, however, take a certain amount of time to do so. An airplane's successful flight depends, therefore, on its ability to move fast enough to offset the effects of the fluidity of the air. A slowly moving wing would displace the air, but air currents would immediately flow into the vacuum created and destroy it. A slowly moving wing would get no support from the air beneath because the air would have time to get out of its way.

The essential of all this analysis is speed. Without speed there is no lift above an airplane's wing and no support below it. Nothing, indeed, except speed can convert the invisible and tenuous air into a substantial support for a heavy machine of steel and wood and fabric. And it will be seen that there can be no amendment or repeal of this principle of flight nor any modification of its laws, unless the air itself is changed, which is a matter beyond the control of aeronautical science.

The essential of flight is motion through the air at a considerable speed. By means of speed the airplane flies in its own element, as the boat floats in water and the wagon rests on land. And to this principle I could find no alternative as I reconsidered the theory of flight from its beginnings, in the presence of an airplane which had failed because it was flown too slowly.

That was the inevitable answer to the first problem in theoretical aeronautics. But another invited investigation. Might there be more than one way to provide and make use of the speed which is essential to the success and safety of a heavier-than-air flying machine?

V

A NEW THEORY OF WINGS

POSSIBLY it appears that my analysis of the essentials of flight made a long journey and arrived nowhere in particular. It seems like a lot of argument and explanation for the sake of a somewhat elementary conclusion. But if you have followed me through it, you will agree that we are now thinking of an aeronautical problem entirely in terms of wings and speed. A great many other matters have engaged the attention of inventors, designers and ambitious dreamers, but it must be concluded from our survey of essentials that wings and speed matter more than any of them.

This process of elimination proves useful in another respect. It leads us to admit that remarkable progress elsewhere may leave the essential problem of aviation untouched. We might agree, for instance, that the perfection of the non-magnetic compass was a splendid contribution to the art and science of aviation, but it has nothing to do with speed and wings. We can admit that the magnificent development of airways throughout the world is a triumph for progressive transportation, but this also is only indirectly related to speed and wings. We may admire endurance flights, the accuracy of instruments, the improvement of weather and radio services, the multiplication of airports and many other evidences of progress in aviation, and still say that they are only incidental to the essentials of mechanical flight as we have analyzed them.

I make the point because it is so important to appreciation of the Autogiro's contribution to the science and industry of aviation. My invention does not deny any of the advances of aeronautical theory, nor does it dispute the credit due to the achievements of the past quarter century. In a thousand details it does not materially differ from the airplane. But it differs profoundly in a single important essential, which I have attempted to define as the basic principle of flight. It differs because it applies in a new way the idea of wings in motion as the essential of the flying machine—the law which permits man to fly in a heavy craft of wood, metal and fabric.

The law does not apply merely to flying machines. It applies as well to kites and gliders, which remain in the air only because the air is in motion with respect to them. No kite will stay in the air in a calm; neither will the thinnest scrap of paper or the lightest feather. The bird flies in obedience to the same law and so does the insect whose wings move so swiftly that the eye cannot follow them. Wings in motion—that is the principle of successful flight.

Following the trail of aviation back to its beginnings, I had reached the point where little was left but this fundamental law, enforced by universal gravitation and the nature of the air itself. The next step was to consider all possible types of wings in motion, hoping to find one better than that which had failed my ill-fated airplane in an emergency.

The obvious beginning was to be made with the

bird, which has been flying about its business since the dawn of time and does it still without benefit of machinery. Could a machine be devised that would imitate the flight of birds, using mechanical power as a substitute for the bird's living strength and muscles?

In theory—yes; in practice—no. The ornithopter—the flying machine that attempts to flap its wings like a bird—has been a very persistent idea and a very consistent failure. Two reasons will explain it, though a dozen others are available. One is that the flying motion of a bird's wing is too complex to be reproduced by machinery wherever weight and simplicity are important. The other is that a bird is more or less alive to the tips of its feathers, whereas a flying machine is only alive where the pilot sits. The operation of aircraft must be confined to a comparatively few controls and adjustments; the wings of a bird are capable of a thousand motions, because the bird is presumably aware of all its nerves and muscles and combines their operation instinctively in the evolutions of flight.

To this must be added the fact that mechanical

power gives man superior strength which cannot be efficiently applied to any bird-like machine. It has never been done; probably it never will be done. There is no need to attempt it if other solutions are available.

A second recognized type of experimental aircraft is the helicopter. The apparent similarity between the practical Autogiro and the experimental helicopter has misled many casual observers to mistake the new craft for "another helicopter," which it is not. But while I was weighing the possibilities in 1919, I was well aware of the formidable fact that no helicopter had been built that would really fly, though the idea was much older than practical aviation itself. Jules Verne described a helicopter in the middle of the 19th century and had it flying all round the world, but only in the pages of romantic fiction. In writing his story he supposed that sufficient power was all that was lacking to make his airship effective as well as attractive. But in the 20th century the power was available and still the helicopter would not fly sufficiently well to be of any practical value.

Two or three reasons will explain for the present

why the helicopter offers so little promise of becoming a successful flying machine. Shortage of power is not one of them. Enough power can be developed by steam or gasoline to lift a locomotive, though if it were done successfully it would still be necessary to devise means to convert vertical into forward flight, which immediately complicates the already formidable mechanical difficulties. But that is by no means the entire problem.

A persistent nuisance to the designer of helicopters is the torque or twist induced by any engine whose power is converted into revolutions of shafts or wheels. If the engine puts power to work in one direction, it thrusts back in the other. In the case of a stationary engine this does not matter, for the engine is braced immovably against it. It does matter in the automobile, though its frame is in contact with the ground through wheels and tires, and so the motor is adjusted to offset its tendency to push the car in a curved path. Out-of-line adjustment of tail surfaces does the same service for the airplane.

But a simple helicopter, consisting only of an engine and a vertical propeller, might conceivably get into the air and still be a totally impractical flying machine, for the reason that the thrust of its motor on the propeller shaft would become counter-thrust on the craft itself. Supposing that it stayed in the air for any time at all, the lifting propeller would soon be spinning in one direction and the helicopter with its pilot in the other—a performance hardly appropriate to the perfect flying machine. A great deal of ingenuity and experiment have been devoted to this problem—and it is still a problem.

Another is the problem of controlled flight under the limitations of the helicopter idea. A simple helicopter, able to lift its own weight but without other devices for directional motion and control, would be no more manageable than a balloon and would be not much better than a balloon. It would go up and come down, provided the wind played no pranks with it, but that is not enough.

So there is no such thing as a simple helicopter as a practical flying machine, and no really successful helicopter of any sort to date. A successful helicopter will be one that offers real competition to the airplane as a means of transportation.

There is no such thing. There was none in 1919 and there is none in 1931.

So two possibilities offered no promise as I surveyed the types of flying machine that had already been tested.

An airplane flies on its wings, not on its propeller, and thereby dodges the difficulties of the helicopter. The propeller is driven by the engine's power to drag the wings through the air. Admitting this as the secret of the airplane's approximate success, I put aside the helicopter idea and considered again the idea of wings in motion. The wing of a flying plane is in motion, of course, but it is a fixed wing and its motion is the motion of the whole airplane as it is pulled or pushed through the air by the propeller. Could I imagine a wing which would have a motion of its own, independently of the forward motion of the whole machine?

I could imagine it, but not as a device that would work. I could imagine, for instance, a pair of wings that could be thrust forward like a pair of oars. That would mean wings in motion—an extra motion, additional to whatever forward movement the craft might have as a whole. Unfortunately, however, the wings would have to come back again for a second stroke, losing whatever was gained by the forward push, or else my imaginary machine would have to throw away its imaginary wings at the end of every thrust and find a new pair for another. Obviously there would be an early limit to this sort of thing.

In order to realize the idea of a continuous forward motion of a lifting surface, plane or wing, independently of the progress of the craft as a whole, there was only one thing to do. The wing must swing in a circle and return to its starting point. And when I realized this, I also realized that I was back at the helicopter again and making no progress toward the Autogiro or anything else.

But a large share of the helicopter's difficulties depends on the fact that its propeller or propellers —which are actually its wings—are supposed to be power-driven. Here was an invitation to further speculation. I began to wonder what would happen if they were not power-driven, but were free to rotate of themselves. All previous ideas had been based on movable wings—flapping surfaces,

THE FIRST OFFICIAL ALTITUDE FLIGHT

Miss Amelia Earhart, at Pitcairn field, 1931.

PAST AND PRESENT

OR

THE FIRST AUTOGIRO AND A MODERN MODEL.

Below: Getafe Airdrome, near Madrid, January 17, 1923.

Above: A "parking ticket." Pilot: James G. Ray.

propellers or oars—driven by mechanical means and a gasoline motor. But this was not a part of the essential theory of flight as I conceived it. The essential was wings in motion and nothing else.

The prospect began to look more interesting. There was plenty of simple experience to show that wings could be put into motion without the direct application of power but simply by passage through the air. A child's toy of paper or celluloid, a set of folded wings on the end of a stick, spins as the child runs with it. A simple propeller turns in the breeze. And it seemed to me that an airplane wing of blades or vanes like those of a windmill would turn freely in flight as the craft was pulled through the air by the propeller.

Was this the answer to my curious quest? I did not know, but I was hopeful. There seemed to be many objections and difficulties involved, but the idea answered the conditions imposed in starting my speculations. It was an idea of wings in motion, different to the idea of a fixed plane moving at high speed in a single direction. If my theory of flight was sound, this thing might work. It might work well or badly, but the simplest essentials of success were present in it somewhere. Surfaces moving at sufficient speed, wings in motion—I reminded myself again and again that these were what was needed for a successful flying machine. And a flying machine with a windmill wing would have them.

What else would it have, supposing that it would fly with sufficient safety and stability?

A popular little toy at one time was a device by which a strip of tin, cut and twisted to a shape like that of a propeller, could be spun high into the air. A ball of the same weight would fall back to the earth at an increasing speed. But the tiny propeller would spin down comparatively slowly, might drift in the breeze to a considerable distance and at last land quite lightly, no matter how far it had fallen.

If a revolving wing would render the same service for a flying machine, it would provide a solution for one of the outstanding and obvious problems of aviation. It would permit the craft to land slowly at a steep angle or even vertically. It would do so without the use of mechanical power, for there is no mechanical power present in the spinning toy propeller. The air makes it turn and because it turns it descends slowly and therefore lands lightly. Would a freely revolving wing do the same for a flying machine?

You will notice that the answer to each question suggested in turn another. But my speculations were now reaching the point where they must be submitted to experiment. Actually, of course, I had questioned my theories much more closely and scientifically than appears in this summary. I had applied to them my knowledge of aerodynamics, which had warned me of certain difficulties and at the same time had given me a degree of certainty that I was on the right track. I had reduced many of my speculations to mathematical certainties and others to probabilities which still awaited the check and proof of experiment. But the two principal points presented in this argument were enough in themselves to justify a trial of my theories by means of models.

One was the belief that flight was possible by means of a revolving surface as well as by a fixed plane, with the possibility that the motion inherent in freely moving wings would reduce the need for a high forward speed of the whole craft. The other was the certainty that a flying craft with a revolving wing would land slowly and more or less vertically. I was still faithful, you see, to the idea that speed and wings are the essentials of flight, and I had imagined a machine with different wings and a different method of providing them with sufficient motion through the air. It remained to discover whether it would work.

Would a freely revolving wing, built like a windmill of blades or vanes, actually turn as the machine went through the air? Would the machine fly on it—that is, would it have "lift" enough to raise the craft and hold it in the air? Would the blades continue to turn at sufficient speed as the machine came down? Could they be built with sufficient strength to be practical? Would the resulting flying machine be stable, manageable and useful?

The only way to find out was to build one and see.

VI

THE UGLY DUCKLINGS

MY first attempts to apply the theory of autogyration to a full-scale flying machine resulted in some curious craft. Compared to the Autogiros of today they were crude and clumsy affairs, designed with no regard to appearance and quite obviously composed of old airplane parts with a rotor and some devices for control. A glance at their pictures reveals, moreover, that one model was not superficially much like another, as though I had experimented with half a dozen types of machine.

That, however, is not strictly the case. Varieties in fuselage and other more obvious features in the early machines are quite unimportant. I made use of what materials I had or could get, building my experimental devices on old airplane frames and using several different engines—most of them second-hand. And though I tried many types, their varieties were just so many approaches to an identical problem. Until this problem was solved it mattered very little what the Autogiro looked like.

At the same time I submit that the crudity of the early Autogiros was no greater than was characteristic of the early airplanes. It seems almost incredible today that the fragile "crates" which were the airplanes of a quarter century ago could have flown. They hardly look like the lineal antecedents of the sleek and graceful airplanes of today, some of which seem to embody the very spirit of speed in their flowing lines and clean contours. The early planes were like great box kites, usually with a precarious perch in front for the pilot and a complicated pattern of struts and wires to support the wing surfaces and controls. Undoubtedly they were ugly, crude and clumsy, as were the earlier members of the Autogiro family.

But an important difference between the primitive airplanes and my ugly duckling Autogiros was that the planes would fly after a fashion and my first Autogiros would not.

During the experimental years and before the Autogiro was ready for general public demonstration or commercial production I built or supervised the construction of some forty or fifty different machines. Many of these, however, represented only slight modifications on their predecessors and there have been only about eight general types. These deserve some discussion; differences in detail among the machines in each series were no more nor less than might be expected in attempting to realize a principle in practice and to perfect the performance of the machine employing it.

The earliest Autogiros, as I have said, would not fly. Two types must be considered as total failures; a third actually left the ground but not in any sort of sustained flight. With the fourth came a radical modification which proved to be the secret of success.

My theoretical studies and analysis had convinced me that flight was possible by means of a freely revolving wing. From the beginning, therefore, I designed the blades of the "windmill," which is today called with more accuracy the rotor, with a proper airfoil. I also knew that this rotor must be free to turn during every moment of flight and that if it failed to do so my machine could not possibly stay in the air. The first problem, therefore, was to build a windmill or rotor of correct proportions, and mount it on a mast structure above a fuselage.

This involved careful aerodynamic calculations as well as some interesting mechanical problems. But in designing and constructing my first Autogiro in full scale I was aware of two more formidable problems involved in flight by means of rotating wings. One was the inevitable obstacle of gyroscopic action, induced by the rapid revolution of a considerable mass around an axis. Everyone who has played with a toy gyroscope knows its characteristic behaviour—that it resists every effort to move it out of the plane in which the wheel is revolving and that any thrust against it is transferred at right angles to its original direction. This promised to make the problem of control rather difficult in any craft flying by means of a revolving wing. Later I discovered, indeed, that it presented an impossibility. Human hands and muscles could not manage or manœuvre an Autogiro if nothing were done to eliminate the gyroscopic effects of whirling the mass of its blades at the speed they would turn.

The other difficulty, discovered in my theoretical calculations as well as by experience, followed from the fact that the blades on one side of the craft would be going forward while those on the other were going backward, so that there would be much more lift on one side of the machine than the other. The difference, indeed, would be very considerable. An average speed of the tip of the rotor blade may be assumed as 200 miles an hour; actually, of course, it varies according to the length of the blade, and the speed the rotor turns. But assuming 200 miles an hour and further assuming a forward flight of 100 miles an hour for the whole craft, the advancing blade would have a net speed through the air of 300 miles an hour, while the receding blade would be travelling at 100 miles an hour. Obviously the advancing blade would be lifting the craft far more effectively than the one that was receding on the opposite side.

My first attempt at constructing an Autogiro proposed to care for this condition by building two sets of revolving blades, one above the other. It was reasonable to suppose that this would eliminate both difficulties if the two rotors turned at the same speed. The experiment has been tried of attaching two gyroscopes together, one above the other, and rotating them in opposite directions. The result is that there is no gyroscopic action whatever, nor will the composite gyroscope stand up and spin like a top. And in regard to lift, this would be completely equalized by two rotors turning in opposite directions at similar speeds.

But unfortunately my first machine would not work. It developed the identical behaviour that I was trying to avoid. As soon as the machine picked up forward speed on the ground and the rotors began to turn, the craft tilted over on its side. It was something of a satisfaction to prove that the rotors would actually turn in the flow of air induced by the forward motion of the machine, but since they would only lift one wheel from the ground they could scarcely be considered as a practical type of wings.

The rotors were identical in size and shape and mounted with equal freedom to revolve at the same speed. But I discovered at once that they did not do so. The lower made only about half the speed of the upper; later I checked my observations and found that the upper rotor turned at about 110 revolutions per minute while the lower was turning at about fifty. The reason, I found, was that the down current of air from the upper rotor interfered seriously with the free flow of air that was supposed to turn the lower blades. The air conditions were entirely different, so the rotors turned at different speeds and my machine had much more lift on one side than the other. And so it would not fly. It would only roll over on its right side.

That was the end of the double-rotor idea for the Autogiro. Its failure makes impracticable if not impossible the suggestion that is still made now and then by observers of the Autogiro—that it might be built with multiple rotors. I presume that the idea is inspired by confusing the Autogiro's rotor with the propeller and imagining a multi-rotored craft to compare with the multi-motored airplane.

The first Autogiro had a rotor placed above an old Deperdussing fuselage, probably dating from about 1911, and was powered with a rotary Gnome engine. Its failure left in doubt the success or otherwise of another idea—a vertical fin above the rotor by which I hoped to secure lateral control in flight. But I never got that far with the first Autogiro.

Neither did I with the second. This was built on a Henriot fuselage with a somewhat more powerful motor. I had definitely abandoned the double-rotor idea and now approached the problem of uneven lift from another angle. This time I tried a three-blade rotor, more in an experimental mood than from any certainty as to whether three or four blades were better. Subsequently, indeed, I tried as few as two and as many as five. Recent experiments lead me to believe that for certain purposes the three-blade rotor may prove the most efficient, but my first three-bladed rotor was in this respect purely experimental.

This time I proposed to overcome the inequality of lift by building into the rotor head a device for changing the angle of incidence of the blades as they made their circle. As they went forward they would be changed so that they would have less incidence; as they retreated the angle would be correspondingly increased, which would equalize the lift on the two sides of the rotor.

My mechanism proved cumbersome and substantially ineffective. The same idea has been tried since, however, by experimenters with helicopters. They are faced with a similar problem and have attempted to solve it with complicated methods of changing the pitch of their lifting propeller blades by some sort of cam action. Unfortunately for their success, this is not the whole of the problem. The correct blade angle for one speed is not the same as for another, so that in actual flight the pilot of such a helicopter would have to be adjusting the angle of his propeller most of the time to allow for the almost constant variations in his forward speed as well as for different angles of descent and ascent. A device has actually been developed to do this, but flying the craft that uses it is an extraordinarily difficult and delicate operation.

In my second Autogiro, if I set my cam action to adjust the blade angle automatically and correctly for flight at 80 miles an hour, it was wrong for 75 miles an hour or for 85 miles an hour. But it turned out that this was only a theoretical difficulty, for the machine never left the ground with more than one wheel at a time. Like its predecessor, it turned over before it started to fly. No doubt this was all for the best, for if it had raised itself into the air it would probably have crashed almost at once.

The third machine was built with a five-blade rotor. This time I attempted to correct the tendency to tip over by means of a system of ailerons, which was partially successful. The machine actually left the ground for a few moments. It could scarcely be called flight; it was more like a hop. But it actually lifted itself clear for a few feet, rewarding its designer with a tremendous thrill and with encouragement to continue his experiments. I doubt, indeed, that I ever found so much excitement and satisfaction in any mechanical success or any flight as in this unsteady performance of my third Autogiro. For it proved the possibility of flight by means of revolving wings, a fundamental principle of the Autogiro and an essential secret of its success.

But there were many things wrong with this model. The most obvious was one that I had expected—the powerful gyroscopic effect of the large rigid rotor. For some time I experimented with controls to overcome it, but with no real success. Something profoundly important was still lacking to the success of my idea, and for lack of it my experiments were more or less futile. I had proved that flight by autogyration was possible; I had not proved that it was practical. The third Autogiro would jump a little, but it would not fly, and sometimes it rolled over like the two that had gone before. And I knew that even if it should fly, it would probably prove impossible to fly it to any useful effect.

At this time I delayed my studies with full-scale machines to test out many theories with little models. Among these was one which behaved remarkably well. It was a four-blade affair, the blades made of thin rattan, powered with the usual twisted rubber cords. It flew steadily and smoothly, the rotor turning according to expectations. Moreover, it would end its brief flights by drifting down to the ground in what might be called a good landing position.

I was puzzled to account for the success of this model compared to my failure to reproduce it on a full scale. Here was exactly what I was seeking, and yet it would only work in miniature. I was aware, of course, that there is a vast difference between flying a model and flying a machine with yourself sitting in the middle of it, but the differences have usually to do with the effectiveness of controls and the smoothness of flight, rough action being imperceptible in a small-scale model. But a much greater difference lay between the little Autogiro with its rattan wings and the machines I had built to full scale. The model would fly; the full-sized craft would not leave the ground and would be uncontrollable if they did.

Invention is more often in debt, I think, to persistence than to inspiration. But it does happen that persistent application to a problem and long

A MODERN CRAFT IN A MODERNISTIC SETTING

The Autogiro of the Detroit News photographed from above.

THE AUTOGIRO OVER MIAMI

vance in design and every incident of experience in developing the Autogiro to its present condition.

The model's rattan blades were flexible in the strict sense of the term, but this did not seem essential to success with a full-scale machine and would have involved great difficulties in construction. Attaching the blades by hinges at the root would serve the purpose. My little model had hinted that if the blades were free to move, the current of air induced by forward motion would adjust them automatically into the correct position for flight.

Since I considered this possibility exhaustively in theory before attempting to put it into practice, it may be appropriate to anticipate a little and briefly describe what actually happens to the Autogiro blades as they turn in forward flight. I have discussed the serious inequality of lift on opposite sides of the craft and my failure to overcome it by mechanisms. But the simplest result of giving the blades complete freedom to move up and down in flight is that they take care of this factor themselves. The forward-moving blade, which has the greater lift, simply rises and so automatically reduces its effective angle of incidence and loses some of its lift; the retreating blade drops, which increases its effective angle of incidence or attack and equalizes its load with that on the other side of the craft. This phenomenon needs more careful discussion, but this will serve for the present. Its effects are easily apparent to observation of an Autogiro in flight; they may also be seen in photographs of the machine in full forward progress in the air. On one side the blades are riding higher than they are on the other.

I began to apply this principle with my fourth type of Autogiro. This had hardly any of the characteristic features of the present machines and the rotor was crude and clumsy. I built it on an old Henriot fuselage, powered with a Le Rhone rotary motor. I proposed to meet the need for controls by tilting the entire rotor assembly from the cockpit by means of a rocker arm. There was no fixed wing, for this was added considerably later in the development of design. The rotor blades were of Eiffel 106 airfoil section, and the rotor disc was 32 feet in diameter.

The articulation of the blades is the term now

used to describe the provision of free movement at the root of the blades or its point of connection with the axis or mast of the rotor. In all the Autogiros built until 1931 cables were attached to the blades to keep them from falling too far, principally because they would otherwise droop to the ground and interfere with other parts of the machine while the rotor was at rest or turning at low speed. During my early experiments I supposed that it was preferable to keep the blades from coning, so that they would remain in a horizontal position while in flight. This was arranged by bending the tubular spars of the blades at the end and attaching them there by hinges, as may be readily understood by study of the picture of the first hinged rotor. Later I found that this had no aerodynamical advantage and involved structural difficulties, so the modern Autogiro flies with its blades lifted a little above the horizontal, or in what may be called a positive cone.

This fourth Autogiro was a very simple machine. Its fuselage and motor were an old airplane assembly. The rotor consisted of four blades built like a narrow airplane wing and hinged so that they could rise or fall freely in the air. They were mounted to turn freely on their axis, and my earlier experiments had proved that they would actually do so as the machine was pulled along by the ordinary airplane propeller in front. Here was the revolving wing of my earliest speculations, with the added characteristic that its four members were free to move in a sort of flapping motion wherever they liked according to the effects of the air upon them. I had considered carefully and even proved to my satisfaction by applying the laws of aerodynamics that this free movement would result in stable flight, which had been impossible with all previous models. But only actual experiment could check my theories and conclusions.

The earliest tests of the new machine suggested that success was coming closer. The rotor turned properly and the machine showed no disposition to turn over on its side as the others had done. Finally the pilot and I were ready to use the full power of the motor and attempt to rise from the ground.

My test pilot was Lieutenant Alejandro Gomez Spencer, a Spanish gentleman whose surname and appearance both indicate an English ancestry. He was and is now one of the best known Spanish fliers and had been instructor for the Spanish Government in Getafe. He was in 1923 a lieutenant of cavalry attached to the air service. It belongs in the history of the Autogiro that he was the first man in the world to fly successfully in the new type of flying machine.

For when we came to the final test the machine lifted at last from the ground, flew steadily across the field and landed safely. This was on January 9, 1923, at Getafe Airdrome, Madrid. A few such flights and we were ready to repeat the performance in the presence of official witnesses. On January 21, 1923, Lieutenant Spencer made a straight flight across the flying field at Getafe before a group of military and Aero Club officials. A little later, at Quatro Vientos Airport near Madrid, he made a four-minute flight over a closed circuit, which was officially certified by Commander Herrera of the Army. On December 12, 1924, J. Loriga completed the first cross-country flight in an Autogiro by flying from Quatro Vientos to Getafe, a distance of only a few miles but a sufficient demonstration of the possibilities of the new type of flying machine.

It is to be remembered that in spite of crudities and uncertainties and a variety of irregularities and discomforts in flight, the behaviour of this primitive Autogiro was fundamentally like that of its most recent descendants. It could fly at a high speed or at a speed much lower than that of an airplane and it landed lightly and came quickly to a standstill. But its principal importance was that it demonstrated beyond doubt two basic principles. One was that the hinged or articulated rotor blade would automatically compensate for varying factors of lift; the other that it would at the same time eliminate all gyroscopic effects. There was still much to be done, but when the fourth Autogiro had made its trial flights and demonstrations success was assured. Theory had been substantiated with proof. And every difficulty that remained was a matter of improvement, refinement and mechanical perfection. The essential principle was established, just as flight by means of fixed wings was proved possible with the first airplane that left the ground. Development thereafter was a matter of research, experiment and patient development.

For nearly three years I built and rebuilt my experimental machines, following the first actual flight. One of the earliest changes was the addition of aileron controls, lacking in the first Autogiro to fly. These were extended from the fuselage on a simple spar. I had not considered at that time the useful possibilities of the small fixed wing characteristic of the modern Autogiro, which serves a variety of purposes. There have been, indeed, many changes in the design and function of this detail of the Autogiro, but at that time they were subordinate to the greater problem of perfecting the rotor.

In turn I designed and built machines with three-bladed and two-bladed rotors. Each revealed new and important elements in the theory of autorotation. Throughout this period, indeed, I was busy testing theory by practice or developing a theory to account for unexpected developments in practice. When an early Autogiro, for example, showed a disposition to rock violently under certain conditions of flight, it was necessary to discover the cause and correct it. There were very many difficulties and even discouragements. Sometimes I went far out of my way to invite trouble, simply because an extreme condition would often reveal an error that was comparatively inconspicuous in a more conservative model. The theory that was formulating progressively with my experiments suggested so many possibilities in this respect that it was essential to certainty to submit some of them to trial.

Most of these models were built on an old Avro 504K fuselage, designed in 1912 and used until the end of the war as a primary trainer, very much as the Curtiss JN4-DC, or "Jennie" was used in the United States. The motor was a Clerget of the rotary type and about 135 horse power. No attempt was made at this time to test the efficiency of the Autogiro in high-powered machines, though I made many theoretical studies in advance of my actual experiments, both in respect to power and size. And during this period I also conducted a long series of wind-tunnel tests with valuable results, among them the determination of the fact that the rotor would continue to turn at every possible angle of flight—a point that was somewhat disputed by critics of my earlier experiments.

At an early point in my studies I adopted an airfoil for the rotor blades which has been consistently used in all subsequent development until recently. Technically this wing-curve is known as the Gottingen 429. Its simplest characteristic is that the curve on top and bottom side is the same. It is a reasonably efficient airfoil, though others give greater lift and a great many different curves are used for different conditions and purposes in designing airplanes. But the important advantage of this particular type is that its centre of lift or pressure is approximately the same at all angles which it may assume in flight. This is not true of other types of airfoil, so that the "centre of pressure travel" is a factor to be reckoned with in using them. Heavy stresses shift back and forth on the chord of the wing of practically every airplane as it changes its angle of attack. This means that the wing must be braced and strengthened to carry them, for they exert a twisting or torsional strain which must be resisted with an ample margin of safety against breaking.

The stresses on the rotor blade of the Autogiro are comparatively simple and are nearly all transferred to its hub or point of attachment as a straight pull. This is easily cared for by structural strength sufficient to the conditions. But a torsional strain on the blades would be highly undesirable. It would necessitate a much heavier construction for the blade itself and for its attachment to the mast. By using a wing curve that has a practically constant centre of pressure under all air conditions, these problems are avoided. So though the shape, size and weight of the rotor blades have varied a great deal during the experimental period, the airfoil has remained approximately the same.

In October, 1925, an Autogiro was demonstrated before officials of the British Air Ministry at Farnborough. This was a four-bladed machine, built on a Standard Avro fuselage and powered with a 100 horse-power Le Rhone engine. The exhibition aroused widespread interest in aviation circles. The Air Ministry ordered a machine for its own use and study, and this was built in England in 1926. A second machine followed, modified in many details and considerably improved in appearance. A third type of English construction developed in 1928, and it was in the following year that intensive development of the Autogiro began in the United States.

A secondary detail in rotor design that came slowly to effective importance during the experimental period was the introduction of another articulation, or hinging, which permitted the blades to move back and forth in a lateral direction, as well as vertically. This wiped out a serious deficiency in the flying of the craft, which had previously been rather rough in flight owing to a sort of whipping action of the rotor blades which jerked at the mast as they turned in their circle. I found that by permitting some movement, automatically determined and controlled by the air currents as are the vertical movements of the blades, this roughness was smoothed out. The movement is imperceptible to observation, but the effect is highly important.

It will be seen, indeed, that the freedom of the blades proved as profoundly important as I suspected when studying the behaviour of my little model with the rattan wings. It permitted the Autogiro to fly, when it would not fly without it. It made it controllable and manœuvrable to the same degree as an airplane and in some respects more so. It banished the bogey of gyroscopic action. And it made the Autogiro comfortable and smooth-sailing, as well as safe and efficient. Real success arrived only when I abandoned all attempt to control the rotor blades and left them at the mercy and guidance of the air currents and their own aerodynamic discretion.

VIII

FLYING ON WINDMILL WINGS

IT was very natural that the public should look around for a nick-name when the Autogiro first appeared in the sky. Obviously the thing was not an airplane, and did not look like one. Those associated with its development insisted on all suitable occasions that it was not a helicopter, though early newspaper accounts of its performances were likely to call it one.

So a number of titles, more or less descriptive, were applied to it. It was described as a "devil's darning-needle," a corkscrew plane, a dragonfly, a flapper flying machine. One imaginative observer thought it looked something like an intoxicated duck; another compared its method of making a turn in the air to Charles Chaplin in his favorite fashion of turning a corner in a hurry. Its thinking about it are sometimes rewarded with a sudden idea which seems to come from nowhere and offer the solution. The important thing is that the inventor shall recognize it when it comes. If he has worked hard enough and long enough and tried and rejected many other possibilities, he is likely to do so.

I was attending the opera in Madrid when I suddenly realized why the model would fly so well and the big machines so badly or not at all. My wife assures me that my excitement at the moment was a little out of place; I doubt that I heard the end of the performance. For it dawned upon me that the construction of the model's rotor blades held the secret of success for the Autogiro. They were not rigid, but built of such flexible material that they would bend easily in flight. In no other respect was there any important or essential difference between the model and the full-sized machines.

It was more or less by accident that I had made the model's blades of rattan; but it was a fortunate accident for my subsequent Autogiro experiments. For automatic adjustment of the rotor blades according to the effects of the air upon them was the secret I was seeking. It had proved impossible to attain by mechanical adjustments, but in the model the flexible wings made their own adjustments and therefore it flew.

VII

THE SECRET OF SUCCESS

IT seems absurdly simple to say that the essential secret of the Autogiro's successful flight is the characteristic of flexibility as applied to its wings. There are many other factors of importance, of course, but it is a fact that success waited on the discovery that a large degree of freedom for the rotor blades was essential to flight and eliminated many obstinate difficulties in other respects. Until this principle was applied to Autogiro design, none of my craft would fly. When I had discovered it, checked it by theoretical aeronautics and applied it mechanically to the machine, the Autogiro took to the air. No subsequent model has failed to fly. The soundness of the principle was made more certain with every ad-

JUAN DE LA CIERVA

Standing by his invention.

TWO EARLY MODELS

Above: First two-bladed Autogiro to fly.
Below: Autogiro built in 1926 for the British Air Ministry.

landing was likened to the alighting of a tired chicken; its take-off to a pheasant rising from cover.

But the name that stuck was that of "the windmill plane." The title is sufficiently descriptive and in some respects accurate, but not in all. It is accurate in that nothing turns the rotor but the wind—the constant wind created by flight or descent. Since this wind is always blowing, the rotor wing of the Autogiro is always turning except when the machine comes to rest on the ground. The Autogiro is therefore a machine that flies on windmill wings. Strictly speaking it is not a windmill plane, for the word "plane" implies the fixed wing and the performance that suits it, but the distinction is somewhat technical. The average layman has learned to call the heavier-than-air flying machine an airplane, without considering closely what was originally meant by the word.

But the windmill comparison is incorrect in an important respect. A windmill's blades are blown around; they do not fly around. The direction of the windmill is determined by the angle of the under side of the blade, that is, the wind strikes against its surface and pushes it around. And the Autogiro's blades actually turn in the opposite direction, since the blades have a small positive incidence.

Autorotation is a fundamentally different phenomenon from windmilling. The rotor blades of the Autogiro are actually flying into the wind, exactly as an airplane flies forward at all times and can do nothing else. It is an accurate explanation, indeed, to say that the four blades of the Autogiro are like four airplanes constantly flying or gliding around a central axis. In technical terms it may be stated thus: that the resultant of the blades' lift and its drag is always forward of the vertical axis. In crude paraphrase of this it may be said that as it goes against the wind the blade is always travelling downhill, or slightly away from the perpendicular axis of the whole machine. It makes no difference at what angle the Autogiro is climbing or flying. The blades are always gliding toward a point a little below the focus of forward flight. It is impossible, therefore, for autorotation to stop while the machine is going anywhere.

This is an absolute fact in theoretical aerodynamics, but it is also an absolute fact of experience. The principle of autorotation has never failed and cannot possibly do so in a correctly designed craft.

In vertical descent the same principle applies; the rotor blades still fly against the wind in their circle. The only obvious difference is that in descent the inequality of lift which causes one blade to rise higher than the other in forward flight is no longer a factor. The blades, therefore, turn at approximately the same level as the machine enters full vertical descent.

So much for the elementary principle of autorotation, which is not merely an empirical matter but has been reduced to exact calculations. By a similar combination of experience and mathematics it can be determined at what speed any specific rotor type will turn.

A second phenomenon that arouses curiosity is the fact that a set of freely hinged blades can support a heavy flying machine. The explanation lies in a very familiar physical fact—the fact of centrifugal force. The disc created by the revolving rotor is actually a very strong and stable flying surface. Considered as a whole it is quite as substantial as the airplane's wing. But its form is maintained substantially by centrifugal force. Turning by autorotation at a considerable speed around the axis, the blades are held powerfully in their position by their own moving weight. The weight of the whole machine pulls them a little out of line, which accounts for their slight cone in flight, but the only other important factor that affects them is the changing force of the air currents upon them, to which they adjust themselves immediately and automatically.

Aviation offers an interesting illustration of the strength inherent in centrifugal force. A German balloon was once designed to fly on a canvas propeller, weighted at the tips. While the motor was at rest, the propeller hung loosely down, so that it seemed impossible that the ship should be able to fly by means of it. But at the high-speed revolution produced by the motor, this canvas propeller assumed the correct shape and held it with sufficient rigidity to propel the airship as though it were made of wood or metal. Its strength and rigidity were produced entirely by its motion, not by its construction.

The Autogiro's rotor blades, therefore, are both bound to turn and bound to maintain their correct flying position. These matters are not determined mechanically or by the pilot's controls, but by laws as invariable as that of gravity. Their effect, moreover, can be closely calculated. It has been determined by long and careful analysis, checked repeatedly in actual experiment, at what speed the rotor will turn, how far the blades will cone up, and what may be expected of rotor blades of different length, weight and design.

The casual criticism of the uninformed observer that the Autogiro's blades seem surprisingly flexible is therefore a fundamental mistake. They are no more flexible in flight than they need to be. Actually it would be dangerous as well as unnecessary to build them too rigidly, for this would deny them the full freedom to make the complex adjustments by which they meet air conditions through which they move. In one of the earlier experimental models, built for the British Air Ministry, too much structural strength was built into the rotor blades. And this was the only Autogiro that ever broke a rotor blade in flight. Incidentally, the pilot was able to land the ship without serious injury to himself.

The wing of the plane must be made strong enough in all its parts to sustain the stresses of flight, by skilful bracing and struts in the biplane and by cantilever or semi-cantilever construction in the monoplane. But in the Autogiro all the stresses are absorbed into and combined with a constant of centrifugal force and transferred into a direct pull or tension directed away from the hub and hinge. It is important, therefore, that the hub shall be made more than adequate to all that is expected of it. There is no more important detail in the entire machine.

The rotor hub is almost entirely a machine shop product. Every possibility of mechanical failure has been considered and provided against. It is made much stronger than it need be, oversize for even ten times the maximum speed of the rotor in flight, and similar care is given to the bearings on which the rotor turns and to their lubrication. The bearings are multiple, and are also mounted to permit rotation of their races in the housings in case of necessity.

So the pilot or passenger in the modern Autogiro may dismiss the three apprehensions that are likely to occur at first acquaintance. The rotor will not stop and cannot stop. The blades are no more likely to break than the steering wheel of a first-class automobile. The hub is as nearly failure-proof as machinery can be made.

The phenomenon of gyroscopic action has been mentioned. It was well known in the days when rotary engines were more common, for these involved definite difficulties in control because of the whirling mass of the motor. It was present in the early and unsuccessful Autogiros, because they were built with fixed rotors. It is sufficient to say that it is completely eliminated by the articulation of the rotor blades, which no longer set up a powerful inherent force which must be overcome in order to control the craft. The Autogiro responds as readily to the controls as the most manœuvrable airplane. If gyroscopic effect were produced by the revolution of the rotor it would be practically impossible to bank the craft for a turn or to bring its nose down or up for a landing. If extraordinarily powerful controls were provided to do so, their action would need to be highly complex, for any force exerted against a gyroscope is immediately diverted to a different direction from its original pressure. But as the Autogiro flies on its free blades, the same air currents which are powerful enough to sustain it are the forces which adjust the wing surfaces of the rotor to the best possible position for flight.

All the useful manœuvres of the airplane are available to the Autogiro. The most undesirable and dangerous ones are not even possible to it. It is quite impossible, in the first place, to stall the Autogiro in the airplane sense of the word. In flying an airplane, the stall is the consequence of losing lift by insufficient speed. The Autogiro blades cannot lose their lift, for the same reason that they cannot stop turning. They cannot be put at a stall angle, for the pilot has no control over them. Whatever advantages are partially provided for the airplane by automatic devices to postpone or prevent the stalling point are present at all times to a one-hundred per cent degree in the Autogiro. It is literally stall-proof.

For similar reasons the Autogiro cannot spin. The spin may be described as a condition in which an airplane is "loose" in the air, turning laterally or vertically as a unit without regard to its controls. In normal flight the unit is the main mass of the machine, and its control surfaces are placed well beyond this mass where they can take hold of the air and secure a sort of leverage by which to move the plane at the pilot's will. But in a spin, an airplane is something like a boat caught in a whirlpool. Everything is moving at once, almost as though the plane were a solid mass or disc.

This behaviour is impossible to the Autogiro because the air foils supporting it are always moving at a high rate of speed. Even in complete vertical descent with idle controls, the lifting surfaces are supporting the craft in a stable manner. The airplane in a spin is not flying but falling; the rotor blades of the Autogiro are flying at every moment that they are moving and they are always moving while the Autogiro is off the ground. The Autogiro, therefore, is spin-proof.

Flying on windmill wings presents one characteristic that is apparent even to the comparatively inexperienced passenger. The familiar "bumps" associated with flying in rough air are largely eliminated, principally because the blades are not rigid surfaces but have a "give-and-take" inherent in their hinged construction. In bad weather, sunny weather or over irregular country the air is frequently by no means a uniform element. Stray currents move up and down and the velocity increases and diminishes. The result is that the wings of a flying machine have sometimes more and sometimes less lift, and in particularly rough air the effect on an airplane is something like the effect of driving fast over a very rough road in an automobile. But because the Autogiro's blades are constantly adjusting themselves to varying air currents, they take care of a large proportion of this variation. The comfort and absence of mental strain is very pleasant to the passenger, who may be entirely aware that bumpy air is not dangerous but knows that it is nevertheless uncomfortable.

An effect of flying in the Autogiro that seems to be largely psychological is the undoubted sense of assurance and confidence it inspires. Those who know something of aeronautics may feel secure because they know how the Autogiro flies and why, and are aware of the margin of safety allowed in its construction. But the totally uninformed layman feels equally comfortable and confident, whether he is riding at full forward speed or coming slowly down to a landing. It has been suggested that this state of mind can be credited to the rotor. The passenger in the airplane instinctively questions every unexpected incident; he wonders if the motor will keep running, if the pilot knows his business and whether the parts of the plane are as substantial as they should be. But the Autogiro passenger naturally puts his faith in the rotor. So long as it turns steadily and smoothly above him, he finds it nearly impossible to worry. And it always does.

Flying on windmill wings is a novel experience to the experienced airplane pilot, but one which he soon learns to relish. To the passenger who has flown a little by airplane it is immediately and unexpectedly pleasant. But already it is beginning to be accepted as the proper way to fly—in comfort, without anxiety and with a complete sense of control and security under all conditions.

In addition, it opens new possibilities of fascination in flight. Those who have travelled much by air know that the keenest pleasure in flying is not found at great heights, where the earth begins to become a monotonous landscape and where the sensation of speed is nearly lacking. What was once called "hedge-hopping" is flying at its best, but in an airplane such flying is rather close to recklessness. But the Autogiro is every whit as safe at a hundred feet from the ground as at a thousand, when over reasonably good flying country. Its pilot and passengers can get a real bird's-eye view of the scene below them, and appreciate all its interest and variety at short range. At will the Autogiro can be flown slowly over interesting terrain or can hover briefly over some point of special appeal; or it can speed at more than a hundred miles an hour to its destination. Its pace can be matched to the mood of its pilot or its passengers.

Flying by means of windmill wings provides, indeed, many new and satisfying sensations. The experienced pilot discovers these in the remarkable controllability of the craft and its wide range of good flying performance. The passenger is aware of increased comfort, a more interesting experience and a sense of security at all times which sets a new standard of enjoyment for air travel.

IX

THE AUTOGIRO IN AMERICA

THE formation of an English company for the development of the Autogiro and the financing of its commercial production resulted in the transfer of experimental work to England. Rights to the Autogiro patents became the property of this organization, known as the Cierva Autogiro Company, Limited. Development was carried on in co-operation with established aircraft manufacturers, who built machines to designs which I prepared with the assistance of a corps of qualified engineers. It was at this time that I learned to fly and became a licensed pilot.

In 1918 a visitor from the United States attended several demonstration flights of the Autogiro. This was Harold F. Pitcairn, at that time president of Pitcairn Aviation, Inc., and a well-known personality in American aeronautics. He had done some experimental work with helicopters and was unusually well qualified to recognize the possibilities of the Autogiro. His increasing interest quickly resulted in the selling by the Cierva Autogiro Company of all American rights, to Harold F. Pitcairn and his organization.

This was the beginning of a close association which has proved highly profitable to the Autogiro and very pleasant to its inventor. It resulted in rapid development of the commercial possibilities of the machine, for the Pitcairn organization was experienced in both the manufacture and the operation of aircraft. Under the title of Pitcairn Aircraft, Inc., it had produced a successful series of planes for mail service, sport flying and primary training. Pitcairn Aviation, Inc., operated a pioneer air mail route between New York, Atlanta and Miami, and achieved an outstanding success in building up the air mail service and patronage in this section of the United States. Associated with this air mail route was a system of four flying schools, in addition to that maintained at the main field near Willow Grove, Pa.

In the summer of 1929 the Pitcairn organization disposed of all its operating activities. It continued the manufacture of the Pitcairn Mailwing and a sport model of the same plane, and began an intensive engineering effort in perfection of the Autogiro. My own work was continued in England, but co-operation was closely maintained across the Atlantic and to a considerable extent the two companies shared the results of their labours. I myself made two visits of some length to the United States. The first was related to the original appearance of the Autogiro in America. The second came during the fall of 1930, when development was well under way in the United States, as was dramatically demonstrated by the fact that I was greeted by four Autogiros flying in formation—the largest number I had seen in flight at one time—as I entered New York harbour on the *S. S. Bremen.*

Much of my time on this first visit was spent in reducing to complete and organized condition the technical theory of the Autogiro and of flight by means of autorotation. This involved some months of work, following the many years in which its

EARLY AUTOGIRO IN FLIGHT

Above: First English built Autogiro for the British Air Ministry, June, 1926.

Below: Autogiro C8, in flight over Newcastle, on a tour of England.

MODERN CRAFT IN A MODERN CITY

Two Autogiros in flight over New York. Note Empire State tower in background.

material had been accumulated. Study and theoretical calculations, checked constantly in free flight experiment, had supplied sufficient data so that a theory could be developed covering many probabilities of performance and possibilities of design beyond the actual achievement in construction to that time. This "first edition" of the theory of the Autogiro forms a large-sized volume, complete with tabulated values and graphs of various items entering into design.

The first Autogiro to fly in the United States arrived there late in 1928. It was flown from the field at Willow Grove, Pa. Mr. Pitcairn had learned with little trouble to fly an Autogiro in England, and one of its first appearances in the American air was when he flew the visiting ship in December, 1928, over his home in Bryn Athyn, Pa. Later he piloted it from near Philadelphia to Langley Field, a distance of several hundred miles, the first cross-country flight by Autogiro in America.

Within two years from its first flights in America, the Autogiro went through many modifications in design, though some of these are only apparent to the technical observer. Rapid progress was stimulated by the intention to enter on commercial production as soon as the remaining weaknesses or uncertainties were eliminated from the machine. No effort was spared to this end, but on the other hand the Autogiro Company of America would not permit any premature introduction of the craft to the commercial market. The work was carried on quietly, in spite of a great deal of public and technical curiosity. Only when we were well satisfied on both sides of the Atlantic that the Autogiro was no longer experimental but a useful, efficient and reliable flying machine were the results of these long years of research offered to the public.

An important American development was the mechanical starter to set the rotor in motion before taking off. The machine cannot fly until the rotor is turning at least 80 revolutions per minute, so it has always been necessary to provide means for getting up rotational speed before attempting flight. In the experimental days this was usually done by starting the rotor with a push by hand and increasing its speed by taxiing around the field.

At one time a mechanical device was developed for the same purpose. A drum was fixed to the rotor, carrying a cable which was fastened at one end to an anchorage on the field. As the machine picked up ground speed and moved away, this cable pulled the rotor around as a string turns a top. But this was an expedient and not a solution to the problem.

Considerable success was secured with a starter which utilized the slip-stream of the propeller to turn the rotor. A biplane tail was built which could be set by the pilot so that the current of air from the propeller was deflected upwards to strike against the rotor blades and turn them. With this device a sufficient speed could be developed in the rotor without running the machine around the ground before taking off. Until this was done the Autogiro was handicapped by the need for plenty of space for the take-off, though it could land in a very small area.

But the mechanical starter proved much more satisfactory, when American engineers developed one of sufficiently light weight so that it did not unduly burden the Autogiro or reduce its useful load too much. The starter now in use on the 300 horse power Autogiros weighs about 65 pounds; the one so far developed for lighter craft weighs about 45 pounds. It operates under the pilot's control, using power from the engine, and it brings the rotor to flying speed in about thirty seconds.

The addition of an efficient starter enables the Autogiro to get out of small fields as well as into them. It could do so before, so far as its length of take-off run and steepness of climb were concerned, but could not do so while it needed considerable space to taxi around in order to put the rotor in motion.

The development of the Autogiro in America received gratifying recognition in March, 1931, when the Department of Commerce awarded it an approved Type Certificate, the highest rating as a flying machine for public use and commercial traffic. It is interesting to note that the Department was obliged to set up new regulations covering the Autogiro, because the terms on which certificates are granted to airplanes do not apply to some of the most characteristic features of Autogiro performance. An airplane is required, for instance, to pass tests demonstrating its ability to recover from a spin. The Autogiro will not stall in the airplane sense of the word, and cannot spin.

A signal honor was conferred on the American organization engaged in Autogiro development by the announcement in April, 1931, of the unanimous decision of the committee in charge that the Collier Trophy should be awarded to Harold F. Pitcairn and his associates. The Collier Trophy is the outstanding American distinction for achievement in aeronautical research for the year of its award. It has been granted to a few individuals, to some governmental departments and to national organizations interested in aeronautics. It was awarded on this occasion for "the greatest achievement in aviation, the value of which has been demonstrated by actual use during the preceding year."

X

PERFORMING IN PUBLIC

FOR the sake of historical record, it will be useful to recall the occasions on which the Autogiro was first presented to public attention and the first demonstrations of its possibilities as an efficient flying craft. It is recognized, of course, that the real record flights are still to be made in the Autogiro. There will be figures set for distance, speed, endurance and altitude. Different types of Autogiro will be developed for special purposes and each will offer opportunities to those who make and break records. But at the time of writing, only six months since the Autogiro reached a sufficient point of perfection to justify commercial production and operation, neither experiment nor competition has determined its performance limits, while it is certain that continuous development in design will permit a series of records to be made and broken before the final figures are determined.

The same was true of the airplane throughout the earlier days of its development. Records which were impressive in 1910, for example, seem now to be rather primitive performances. It was something of a sensation when a plane first flew at one hundred miles an hour or to a height of ten thousand feet, but these figures are now commonplace. Earlier performances are only interesting today for their historical significance.

The first Autogiro flights have already been mentioned. These took place in January, 1923, at Getafe Aerodrome, near Madrid. The third demonstration was the short cross-country flight of December 12, 1924. From this time on the Autogiro was flown regularly in practice and experimental flights, steadily lengthening its range and revealing its possibilities.

On June 24, 1925, the first demonstration flight was made before King Alfonso of Spain at Cuatro

Vientos Aerodrome. The King showed the keenest interest in this flight and followed my further experiments with a friendly and encouraging attention.

On October 20, 1925, the first demonstration flight was made before officials of the British Air Ministry at Farnborough, England, and led to the building of two Autogiros for the Government. February 2, 1926, saw the first flight in France, at Ville Coublay. On July 2, 1926, the Autogiro took part for the first time in an air show, when it appeared at the annual aviation pageant at Hendon, England.

On August 2, 1926, I was the first passenger to fly in the Autogiro, a distinction which I am proud to acknowledge. It may be added that my wife was the first woman in the world to be an Autogiro passenger.

In the same year, on September 5, the Autogiro was introduced to Germany at Tempelhof Field, Berlin, flown by the British pilot, F. Courtney. I flew it myself in its first demonstration in Belgium, taking it by air from Paris to Brussels on October 12, 1928. A special machine built for the Italian government, the C.8.111, was flown at Rome in January, 1929, with myself as pilot.

A more spectacular demonstration was the flight made in an Autogiro by Arthur C. Rawson in July–August, 1928, when the machine made a circuit of England, landing at many airports around the country. A little later, on September 18, it crossed the Channel for the first time. I acted as pilot and carried as passenger the editor of *L'Aeronautique,* M. Bouché. There was a general impression even then that the Autogiro could not fly at a pace to compare to the airplane, but I had the satisfaction of compelling an airplane that accompanied us to do its best to keep up with us. This flight was from London to Paris, and probably attracted more attention than any preceding performances. The newspaper accounts of this journey are interesting but not highly accurate. The Autogiro was still such a novelty that reporters and editors did not grasp its principle, and it was generally referred to as a "helicopter." One dispatch published in an American newspaper at the time called it, indeed, an "autogyroscope." An English correspondent made an extraordinary effort to explain the machine and achieved the following: "The new principle of substitution of a horizontal lifting screw for wings may lead to construction of a reversible plane. With proper variations in its present form, the plane could be fitted with fore and aft rudders allowing both forward and backward flight or an intermediate hovering motion." This, it may be said, is a great deal more than its inventor has ever claimed for the Autogiro or expected of it.

Following these introductory flights in various countries of Europe, the Autogiro began to add mileage to its record in impressive fashion. There were many flights of great local interest, like those that marked the general introduction of the airplane to public attention in its earlier days. The demonstrations in the United States are more or less typical of what went on in many countries during 1929 and 1930, and serve to illustrate the increasing range of the Autogiro and the reception accorded it by the public and the aviation industry.

In addition to such public demonstrations as are deserving of record, the Autogiro was flying steadily in connection with experimental work during this period. It is a conservative estimate, indeed, that by the end of 1930 the machine had been two thousand hours in the air in America alone.

The first flight in the United States was in early December, 1928. The Autogiro used was one built in England, an improved model of the C.8 type, which was merely an old Avro 504-K plane with a rotor instead of wings. A Wright J5 Whirlwind motor had been sent over from the United States and mounted in this machine. The first American flights were conducted privately at Pitcairn Field, near Willow Grove, Pa., but news of them appeared at once in the press and aroused widespread curiosity and interest in aeronautical circles.

A little later comparative flights were made with an up-to-date plane of the period, to discover relative rates and angles of climb, flying speed, and landing roll. The results were encouraging, though contrasts were by no means so marked as in subsequent comparisons with more modern types of Autogiro. But the Autogiro bettered the performance of the plane by about twenty per cent and this was satisfactory enough so that the Autogiro was introduced during the year to the aeronautical and general public by appearance at several air shows.

The G-AAKY, an Autogiro built in England, made the first public demonstration in the United States in August, 1929, at the National Air Races at Cleveland. Scepticism was the mood of the moment in regard to the new craft, so that when the first demonstration was scheduled the announcer remarked through the loud-speakers on the field that "the Autogiro will now take off and fly—if it can." It could and did, and public interest responded readily to the performance.

The performance possibilities of this model were rather limited, but at the next series of races, at Chicago in August and September, 1930, two American-built Autogiros and another English model were flown daily. It may be stated now that the demonstrations at Chicago permitted only an approximate idea of the Autogiro's possibilities. It is estimated that the machines shown at that time could climb at about 700 feet per minute, while the 300-horsepower Autogiro of 1931 climbs at about 1800 feet per minute. Moreover, the Chicago show was held in consistently calm weather which somewhat penalized the Autogiro's performance in respect to the spectacular. A slight breeze—so slight that the spectator scarcely considers it a wind—permits a true vertical landing, which invariably arouses an audience to enthusiastic appreciation.

Following this appearance, Autogiros began to appear at air shows throughout the Eastern United States. At Newark Airport in 1930, the Autogiro was invited to lead the "air parade" and had an opportunity to disprove the prevailing impression that it could not keep the pace of modern aircraft. A little later a special demonstration was given at the request of Mr. Thomas Edison. The Autogiro was put through its paces for the venerable inventor, who expressed his approval promptly and to the point. "That's the answer, that's the answer," he exclaimed, and went on to describe the Autogiro as "the greatest advance that could have been made in aviation."

In the winter of 1930, at Miami, Fla., the Autogiro gave the first public exhibition of its ability to get in and out of close quarters in a modern city. Pilot James G. Ray landed it in a small park, not far from the center of Miami. A touch of true modernity was added to this demonstration when a police officer handed the pilot a "ticket" for illegal parking, but was persuaded to withdraw the charge when the Autogiro "moved on" with the officer as passenger.

The same point was proved in rather dramatic fashion when officials of the Autogiro Company of America went to Washington on April 22, 1931, to receive the Collier Trophy at the hands of President Hoover. As an incident in the program of the occasion, Pilot Ray brought an Autogiro to a perfect landing on the south lawn of the White House.

There will, of course, be many more demonstrations of interest and importance as the Autogiro comes into general use and appreciation of its possibilities. Those in charge of development in America have declined to exploit its spectacular possibilities, and have conducted public demonstrations only to the degree warranted by its progressive development as a practical and tested craft for its proper purpose. This policy was indirectly recognized by the high honor accorded to the Autogiro by the award of the Collier Trophy.

The speech of President Hoover on this occasion is therefore of special significance. He said:

"The invention of the Autogiro by Juan de la Cierva is one of the outstanding improvements in heavier-than-air craft. Its ability to arise and descend with safety almost vertically makes it a practical and decided step forward.

"Six years ago Mr. Pitcairn recognized its value and later brought it to the United States, where he and his associates have continually developed the device.

"By widespread demonstrations they have inspired public confidence to the point where the National Aeronautic Association felt justified in awarding it the Collier Trophy 'for the greatest achievement in aviation in America, the value of which has been demonstrated by actual use during the preceding year.'

"On behalf of the National Aeronautic Association, it gives me great pleasure to present to you and your associates the Collier Trophy for your development and demonstration of the practicability of the Autogiro in the United States. This trophy is emblematic of the highest award in American aeronautics. I congratulate you."

In acknowledgment, Mr. Harold F. Pitcairn, president of the Autogiro Company of America, replied as follows:

"Mr. President: In accepting the Collier Trophy Award for 1930 my associates and I desire to share this honor with Juan de la Cierva, the inventor of the Autogiro, whose genius and perseverance made possible our part in its development.

"It is our firm conviction that the continued development of the Autogiro will lead to the general adoption by the public of flying for both utility and pleasure. We shall continue our work to this end with a profound sense of the responsibility implied by this public recognition.

"We are deeply grateful to the National Aeronautic Association for the distinction they have conferred upon us and we wish to thank you, Mr. President, for honoring us by personally presenting this Award."

A HYDROGIRO

A 1926 model Autogiro fitted with floats.

A CLOSED CABIN AUTOGIRO

Le Pere all-metal cabin Autogiro. Flown in the U. S. September, 1929.

XI

FOUR-WHEEL BRAKES

MANY times I have been asked the difference between flying an Autogiro and an airplane.

The answer might be phrased in a single sentence. The Autogiro compares with the airplane as an automobile with four-wheel brakes compares with one without any brakes at all.

I mean by this that in operating an Autogiro the pilot's control is complete and effective throughout the entire range of flying speed, from nothing to the maximum allowed by the engine. In flying an airplane, control is effective within a much narrower range. It is complete at full speed, but diminishes rapidly as minimum speed is approached. Under certain conditions it vanishes entirely. Such conditions result in the stall or spin, either of which is essentially a failure of control. Unless the pilot can recover effective command of the craft, which can only be done where there is space to spare, his plane falls out of the sky nearly as rapidly as though it were so much dead weight. If loss of control occurs close to the ground, a crash is nearly inevitable.

In a stall, the plane loses its effective lift, another variation of loss of control. It loses its "hold" on the air, without which it is no longer a flying machine. This is not necessarily the result of engine failure, and it should be clearly understood that the term "stall" in aeronautics has a different meaning than when it is used to describe the misbehaviour of an automobile motor.

A stalled motor is simply a stopped motor, the stall usually resulting from improper manipulation of the fuel control. But an airplane is said to stall when the correct flow of air over the wing is too seriously disturbed to maintain lift. First the plane reaches the "burble" point, a phrase descriptive of a confusion of air currents above the wing.

Unless corrected at once, this condition is likely to grow worse and become the stall. The stall may be induced in several ways, commonly by climbing too steeply at insufficient speed or by an incorrect angle of attack in forward flight. Engine failure invites a stall when the plane is climbing, because speed drops almost immediately below the minimum of safety. Therefore the pilot is most concerned for his engine while he is taking off and attaining altitude. If his motor dies while he is high in the air, he can maintain control by manipulating his glide to the ground. But if the power fails on a climb, the plane may stall before it can be brought to a good gliding angle. A pilot may lose control at good altitude and have time and space to recover it. But in a stall or spin below a few hundred feet the best plane and the best pilot may meet disaster.

The faster the plane, the greater the altitude that is needed for recovery from a stall. Particularly the plane with a high landing speed is in serious danger if stalled at low altitudes, for its normally long glide is converted into proportionately rapid descent if the craft goes out of control. Light-loaded planes with low landing speed are safer in this respect and it is sometimes claimed that they are stall-proof. But any airplane can be stalled by bad flying, and if this happens near the ground the results are nearly sure to be serious.

We may fairly compare this behavior with the operation of a brakeless automobile. On a clear road, in good hands, and in the absence of emergency conditions an automobile can be driven without the use of brakes. By manipulating the throttle and taking advantage of the slope of the road, a good driver could stop a brakeless automobile under favouring conditions, could control his speed within certain limits and could escape serious trouble by avoiding it. But an automobile without brakes, running down a steep hill toward a sharp turn in the road or a traffic jam, would be as badly off as an airplane in a spin or stall. It would be hard to handle on a busy road, demanding extraordinary skill and vigilance from its driver. It would be difficult to halt at an exact spot, unless the stop was considered and calculated well in advance. In sudden emergency, the driver could not avoid collision or escape its consequences.

No sensible person drives an automobile on the highways without good brakes. But the airplane pilot has no choice in the matter. Therefore he must compensate for the lack of mechanical means of safety by his skill and experience in flying. He must be constantly alert, watchful of his motor, sensitively aware of his controls and his craft's responses, and wary of dangerous conditions in the air and on the ground.

The same sort of vigilance would enable a driver to use an automobile without brakes in comparative safety, on a familiar highway not too crowded with competing traffic. But such skill is unnecessary, or else there would be very few automobiles on the roads. The invention of good mechanical brakes made the automobile a practical vehicle. Their perfection has justified the rising pace of highway traffic, which was possible years ago but could not be permitted without the general use of powerful and reliable brakes. Without brakes, twenty miles an hour would be a dangerous speed for motorized traffic. With the four-wheel brakes of today, fifty miles an hour is actually a safe speed for the open highway, for it can be reduced at will and in a moment or two to much less,—to nothing, if need be.

Various devices have been invented and tested to provide a sort of aerial brake for the airplane, or to make such a brake unnecessary. Wing slots and flaps somewhat extend the margin of safety from stall. Slots operate automatically when the plane's wings are at a dangerous angle, maintaining the lift beyond the usual limits of an efficient wing. They do not prevent the spin nor much reduce the difficulties of landing. Automatic stabilizers somewhat reduce the pilot's responsibility in normal flight and it is possible with exceptional skill to operate a plane entirely by instruments from a "blind" cockpit under suitable conditions. These devices reduce the danger of human error, but they do not affect the plane itself. It is still a high-speed flying machine without brakes.

About the only braking device that is measurably effective is the large-scale parachute intended to lower the entire plane at something less than a deadly speed. This has been demonstrated, but has not yet been proved practical. To equip all airplanes with so bulky a safety device would be something like providing every rowboat with a raft. Such a parachute, moreover, is a precaution but not a preventive of accidents nor will it help if trouble develops near the ground where the danger of disaster is greatest. For comparison it may be suggested that automobiles carry bumpers, but that these do not make brakes superfluous.

Other braking devices that have been suggested for airplanes are more or less fantastic. At best they patch up a weakness without curing it; at worst they complicate aircraft so seriously as to rob them of their efficiency. Certain devices have been developed which correct one fault but intensify another. The successful and efficient airplane is a compromise among many desirable characteristics. If one is emphasized, another is sacrificed. An unusually low landing speed, for example, must be paid for with wing area or carrying capacity or the loss of useful flight characteristics.

It appears, indeed, to be impossible to put a practical brake on an airplane, by which the pilot can control his air speed at will or stop when and where he wants to. And until an airplane can be stopped at will under all flying or landing conditions, it falls far short of the efficiency and security expected of the automobile, without which the truck, bus or private car would never have become of universal service to transportation. The modern airplane flies marvellously well, but it still lacks its four-wheel brakes.

There are three circumstances under which an automobile needs its brakes. One is in making a sudden turn, a second in travelling a bad road or in bad weather, a third in coming to a stop. In three similar respects the Autogiro is so different from the airplane that comparison scarcely tells the whole story. The superiority is positive, not comparative.

Those familiar with flying as passenger in a plane note several marked differences in riding in the Autogiro. One is that the Autogiro can make a slow turn in mid-air without losing altitude and with no danger from a skid or side-slip. There is a minimum speed—actually a pretty fast pace—at which the plane must turn in order to remain under complete control. At less than this minimum the plane begins to "slide down the sky" and eventually may fall into a stall or spin. But the Autogiro can safely turn corners at slow speed. For the same reason it can make a surprisingly sharp curve in changing its course.

The practical effect of this is that the Autogiro can avoid obstructions, stop short of them, or climb over them in a manner beyond the ability of any airplane of the same power and weight. It can actually be "pulled up short" in the air, exactly as an automobile can be checked by its brakes.

The second advantage of brakes to the automobile is the possibility they provide of driving safely in bad weather. The driver need not be able to see half a mile ahead if he can stop his car easily within a hundred feet.

There are various air conditions which can be classified as bad weather for flying an airplane. Mist, rain, fog and darkness all make flying perilous. A pilot does not attempt to fly in the face of them unless it is absolutely necessary. The most obvious reason is that he cannot see where he is going in time to stop before he runs into trouble. His plane may be flying perfectly, but it must fly at fifty to eighty miles an hour or more, and that is much too fast for safety and confidence in finding a way through bad weather, particularly over unfamiliar country. Nor can he stop in emergency or even check his speed when in doubt what to do or where to go. He has no brakes.

It is not suggested that the Autogiro has any business to be flying under really bad weather conditions or in thick fog. Aircraft cannot move under severe weather conditions, just as shipping is tied up by a really heavy fog and the automobile stands still when the roads are piled with snow. But if an Autogiro encounters bad weather or low visibility in flight, the pilot can fly steadily at thirty-five miles per hour, check his speed completely in a moment, choose a landing place at leisure and land safely. He can even fly below low clouds at a level admittedly dangerous to the airplane, loafing along at comparatively little speed. If he suddenly finds himself in a low-hanging cloud he can safely drop below it and look around, aware that his craft is under complete control though it is no longer making any considerable forward speed.

This parallels the behaviour of the automobile

driver under emergency conditions of road or weather. When in doubt or difficulty he uses his brakes. The airplane has no brakes; the Autogiro's revolving blades are a braking device in themselves, thoroughly efficient under all conditions of flight.

The third service of four-wheel brakes is to bring the automobile to a stop at a determined place. It may be necessary in an emergency; more commonly it is just a matter of halting the car at a convenient location. In either case the driver trusts to his brakes much more than he relies on his engine or skill in handling steering wheel and throttle. The use of the brake simplifies enormously what would otherwise be a difficult and delicate operation.

The Autogiro does the same for the risky and responsible feat of landing a flying machine on a predetermined spot, under nearly any conditions that are encountered in conservative flying. It has been demonstrated again and again in "landing-to-the-mark" contests that the Autogiro in skilful hands does not merely come close to the target,—it lands exactly on top of it and there it stays.

Its descent is so completely under control at all times that its performance closely parallels that of the automobile under control of its four-wheel brakes.

Two mechanical advantages are largely responsible for this continuous control from full flying speed to a standstill. One is the steady revolution of the rotor, which revolves at constant speed so long as the craft is in the air. At maximum forward speed or in vertical descent it continues to make more than a hundred revolutions per minute, kept in motion by forward speed or the uprush of air and kept in correct position by centrifugal force. Neither of these factors, it should be noted, have anything to do with mechanical control or the pilot's skill. They operate entirely automatically. Their result is that the Autogiro retains a true flying speed, though less than what is needed for forward flight, even while it is descending to a landing. There is no possibility of stall. The rotor blades, turning at their automatically determined speed, will not lose their "hold" on the air, as does the airplane wing when the minimum is passed.

The second important advantage is that the down wash from the rotor contributes to the effectiveness of all the control surfaces. It is obvious that the controls of aircraft are entirely helpless if there is no relative motion of the air against them. The airplane keeps its controls effective by rapid forward motion. The descending Autogiro may be so guided as to have little forward motion or none at all, but its controls remain in command because the air around them is kept moving by the rotor. This is an elementary statement of a fact that can best be proved by experience. Stated in aerodynamical terms it would be highly complicated, but it is a simple matter to discover in flying an Autogiro that the controls are still working as the craft comes in to a landing.

Bringing an airplane to a stop, particularly under adverse conditions, is the most difficult manœuvre in ordinary flying. It is nearly the simplest in operating the Autogiro, exactly as it is a small problem to stop an automobile if the brakes are good and reliable. Similar ease replaces difficulty in other essentials of flying, and for similar reasons.

That is why I describe the differences between flying an Autogiro and an airplane as the contrast between operating an automobile with and without four-wheel brakes.

XII

HAPPY LANDINGS

WHEN aviation entered its years of vigorous development, and commercial enterprise began to discover many practical possibilities for the new means of transportation, flying schools began to multiply more or less as a matter of course. There would obviously be a real need for trained pilots if commercial air transportation developed according to expectations, and aviation seemed to offer a promising career to a multitude of ambitious boys and some ambitious girls. They came to the flying schools in surprising numbers, and among them were many of today's famous pilots and makers and breakers of aeronautical records.

It was only by bitter experience that some of these ambitious youngsters discovered that much more than a preliminary course in flying is needed to achieve a career in aviation. They learned that a ten-hour course is only an elementary apprenticeship to the profession of pilot; that a school can give training but not enough experience to make a first-class flier; that it is a long and expensive process to prepare for the responsibilities of a real job in aviation. Many of them found themselves at the end of their funds when they were still far from sufficient proficiency to compete with the veterans flying the air mail and the transport planes. Many of them quit, disappointed and discouraged with the failure of their plans and purposes.

There was a time when the shortage of competent pilots was so serious that the schools were justified in holding out hopes of a profitable career at the end of a flying school course. The growing industry was willing to take the graduates, school them further in active service, and put them to real work when they were ready. But this did not last long. The student-pilots multiplied much more rapidly than the planes in use, so that there were a dozen hopeful candidates for every seat in a cockpit. Moreover, the industry was compelled to tighten

A NEW CRAFT AND A NEW BRIDGE

Two Autogiros flying over the new bridge across the Hudson River, September, 1930.

THE WINGS OF TOMORROW

Six Autogiros flying in formation at Pitcairn Field, Pa.

up on its requirements. The risks and responsibilities of the pilot's profession were more closely analyzed, and operators demanded long experience as well as a license in choosing and hiring their men. Governmental authorities adopted similarly strict standards. In the United States, for example, the Department of Commerce several times raised the requirements of its licensees, until the ten hours of the average flying school seemed a very small part of the two hundred required for a transport rating. And in actual competition for the better opportunities in the pilot's profession, several hundred and even thousands of hours of flying experience might be part of the veteran's credentials—a formidable handicap to the hopeful but inexperienced youngster.

At the end of 1930, after two years of trouble for the aviation industry, the well-disposed experts were warning the younger generation away from the flying schools. Such a warning was needed, for the boys and girls of today want to fly. Their typical interests and impulses are satisfied by mastery of the plane's machinery and the fascination of flight. And it is significant, though not surprising, that an extensive educational survey, conducted by professors of the University of Michigan in 1930, discovered aviation as the outstanding ambition of boys between the ages of eight and eighteen. But there were not nearly enough opportunities in aviation at that time to take care of a fraction of these fond hopes and energetic ambitions.

There are three obvious reasons for the discouraging prospect facing the student pilot at the end of aviation's first quarter century. One is the scarcity of planes in need of commercial pilots. A second is the extraordinary amount of training and experience needed for real proficiency. The third is the improbability of the average young man becoming the owner and operator of his own aircraft.

In these respects the Autogiro will fundamentally affect the situation for the better. It is, for many reasons, the flying machine of the future for the private owner. Its maintenance and operating costs are lower than those of the plane, because it is much less liable to expensive damage from ordinary accidents and because it can be flown with considerable independence of the commercial airport. For the same reason it has a wider range of usefulness for the private owner. He can use it for sport, business and travel, not merely on the established airways but nearly anywhere.

But there is much more encouragement for flying students in the fact that it is easy to learn to fly the Autogiro with sufficient confidence and security for useful service and real enjoyment. Comparisons have been made by experts in operating both the airplane and the Autogiro. It is estimated that ten hours in the air are the average requirement for minimum ability in flying and landing an airplane. At least fifty are needed to acquire sufficient skill to cope with emergency conditions and to operate a plane commercially or with any degree of confident enjoyment. Several hundred hours in the air are a minimum of experience without which a pilot should not attempt the transport of passengers or valuable cargo.

It is a conservative estimate that an intelligent student will be ready for solo flight in the Autogiro after two or three hours instruction. At the end of ten hours' practice he will be able to handle his machine sufficiently well to use it freely for pleasure. Whatever further flying experience he may obtain will, of course, make him a better Autogiro pilot. But the important contrast is that at the end of ten hours the Autogiro student is ready to fly without anxiety to himself, while ten hours is nothing but a beginning for the airplane amateur.

The point of proficiency—enough for ordinary purposes—is reached in the Autogiro in about one-fourth the time required for the airplane. This is important on two counts. It means that the cost of owning and operating a flying machine, which must include the cost of learning to fly it, is very considerably reduced, so that private ownership is stimulated on another score. It means also that the expense of real flight training will come down to much lower levels when Autogiros are in general commercial use. The would-be pilot will not need to spend as much time and money as a student doctor or lawyer before he is considered safe to fly about his personal business or for the fun of it.

For these facts there are perfectly practical and demonstrable reasons. Most of them have to do with landing.

Under ideal air conditions it is not difficult to fly an airplane, and the student pilot soon masters the trick of it. It takes him much longer to learn how to put his plane down on the ground in a good landing. The solo point in his training, in fact, is not determined by his ability to manage the controls in flight but by his ability to bring the plane back to earth when the flight is over. Those who have watched a flying school at work will have seen the students taking off and landing repeatedly under the advice of their instructors, learning to start and stop.

Every student knows the difficulty of landing and the flying public is more or less aware of it. Visitors to airports approve the skill with which the experienced pilot makes a perfect three-point landing; passengers in transport planes comment on the smoothness and accuracy with which their heavy craft are brought to earth. It is accepted evidence of flying skill, indeed, to be able to make good landings under all conditions encountered in the operation of a plane.

It is a fairly complex problem that faces the student as he learns to land his plane. He must estimate the maximum and minimum gliding angle of his craft and select a spot within their range for his landing. He must calculate to clear all obstructions around the flying field. He must discover the wind direction and land against it or allow for drift in a cross-wind. He must leave room for his landing roll, or apply his wheel brakes smoothly to check his speed on the ground. He must learn to look all around him as he comes in to the field, and to estimate the condition of the surface in an unfamiliar landing place. He must practice until he can consider these factors quickly and decide at once what to do about them.

But more important still is his ability to keep safely within the limits of a proper landing speed. Too fast is bad; too slow is likely to be much worse. By means of his instruments and the "feel" of the ship he must come as close as possible to the minimum landing speed without running the risk of dropping below it. This is undoubtedly the student's most critical problem. Failure to figure it correctly has caused more minor damage to student planes than any other error of inexperience. Smashed undercarriages, "pancaked" planes, wings crumpled and a variety of other expensive accidents are likely to be the price of learning to keep on the safe side of a minimum landing speed. And sometimes, of course, the consequences of error are much more serious.

A clumsy landing or even a crash may also be caused by psychological failure, rather than an actual error in judgment. The student has so much to think about that he loses his nerve just when he needs it most. In his confusion he either makes a mess of his landing or tries to climb out of his difficulties and runs the risk of a stall, just where a stall is most dangerous.

Most of these difficulties cannot be entirely avoided in learning to fly a plane, though they may be minimized by confining practice to an ample flying field and a type of plane that is deliberately designed for simplicity and stability. But at best it is no easy matter to learn to fly an airplane. And the greatest of its difficulties is to learn how to land.

So the advent of the Autogiro is profoundly important to the future of flying and the problem of those who wish to learn to fly. For the Autogiro has made a comparatively simple manœuvre of the student's chief difficulty, and reduced in proportion the responsibility of his instructor. Each of the operations and calculations involved in landing an airplane is immensely simplified in landing the Autogiro. The range between maximum and minimum gliding angle is greatly increased, so that the flier need not aim so accurately as he approaches his field. The ability to make a vertical descent eliminates anxiety concerning obstructions, and there is no need to allow for landing roll or to worry over the possibility of colliding with objects on the field while slowing down to a standstill. Much more important is the fact that the student has no need to consider or calculate either maximum or minimum landing speeds. In vertical descent his maximum rate of fall is fixed automatically by the revolution of the rotor, so that if he forgets everything he has learned about landing he will still come down slowly, without hurting himself and probably without damage to his ship. On the other hand, the Autogiro has no minimum landing speed short of no speed at all. The student cannot miscalculate this factor, nor does he lose control at low air speeds. If he allows his pace to drop below what is necessary for forward flight or if his engine stops, the Autogiro simply settles to the ground, requiring little guidance to do so in good landing position.

The psychological hazard is also immensely reduced in operating the Autogiro. A student in difficulties or confused by conflicting ideas as to what to do next can make a landing with ease under conditions which would call for first-class skill in landing an airplane. If he discovers, for instance, an obstruction looming in his path of flight he need not risk a stall in climbing above it; he can simply come down in front of it. An airplane pilot in a similar fix must either pull his ship sharply into the sky or make an extraordinarily skilful landing, taking a chance meanwhile of a stall or spin at a dangerously small distance from the ground. Since the Autogiro will not spin nor stall, even under amateur handling, a persistent anxiety of the student pilot is eliminated from his practice periods.

The inherent characteristics of the Autogiro contribute in other respects to the student's peace of mind and progress, particularly by eliminating the danger of falling off in a spin or stalling under any conditions of flight or motor performance. But the simplification of the technique of landing is typical of its advantages for the flying novice.

There is, of course, no such thing as a fool-proof flying machine. There is no mechanical cure for clumsiness and raw students will now and then upset an Autogiro on the ground or otherwise cause trouble for themselves and their instructors. Mistakes will be made, but they are unlikely to lead to serious consequences. There will be accidents, but nobody is likely to be hurt.

Undoubtedly the general use of the Autogiro is going to reduce the expense of learning to fly, cut down the high cost of student crack-ups and convert the high-priced profession of the pilot into the useful trade of the flier. It will mean that flying school will be a succession of happy landings, instead of a series of anxious moments for pupil and teacher. The Autogiro student will be able to take it easy as he learns to fly. And he will be able to fly to some useful effect when his elementary training is over.

XIII

THE FAMILY FLYING MACHINE

AVIATION in America and elsewhere did not amount to much before the Great War. But under the extraordinary competition of international strife, designers worked with a sort of desperation to improve their product and factories turned out airplanes without much regard for cost efficiency. Planes were literally wasted by hundreds, types became obsolete overnight, engines and parts were produced by thousands and money was spent by millions. Back of the fighting lines there grew up in most countries an industry tuned entirely to war and the needs of war. And within four years aeronautical science had made more remarkable progress than in all its previous history and was still going strong when the war ended.

But when it was over, aviation contributed its full share to the scrap heaps of the nations, piled high with machines and materials made for war and of no apparent use for peace. It was not merely a matter of surplus army stores, factory stocks and glutted warehouses behind the empty trenches. These were bad enough, part of the fearful waste of modern warfare and preparation for it. Americans will remember many examples of it, including such absurdities as the sale of great propellers, which had once cost the Government many hundreds of dollars, for less than a dollar apiece. But beside the obsolete war material, there were the factories in America and elsewhere which had been built to provide and maintain the world's aerial fleets. There were skilled workmen and expert designers with a new wealth of aeronautical knowledge, made idle by the coming of peace. There was an entire industry, developed to the scale of war and left stranded in a world that had not learned to fly on peaceful errands.

Aviation did not, of course, fold up its wings and wait for another war. It turned hopefully to the uses of peace, accepting the war as an incident in its extraordinary history which had made it grow much faster than had been expected when the baby science and industry were born. As soon as the war was safely out of the way, aviation looked around for something to do.

The aviation industry decided to go into business, very much as the "tank" factories went back to building tractors and other farm machinery. But there was a difference. Farming was a familiar occupation long before the war; flying was not. Aviation could not convert its post-war energies directly to supplying the needs of commercial air traffic, for there was no such traffic. Its problem was not merely to supply the needs of an established industry, but it was also compelled to create it.

So there began soon after November, 1918, an immense effort in all industrial nations to establish commercial air traffic, a new type of transportation utility. The effort still continues, successfully and otherwise, according to local conditions. It is particularly vigorous in America, where the intelligent co-operation of the Government by means of the air mail program has produced more real progress than any of the direct subsidy systems of Europe.

But the remarkable result of this intensive effort has been that two types of airplane dominate the skies today. One is the military machine, far more deadly than any which flew in France in 1918. The other is the frankly commercial aircraft, designed and operated to yield a profit by carrying passengers, mail or light freight. The proportion of flying done today by other types of plane is negligible.

There are very few privately-owned airplanes, compared to the number of those that are designed and operated for commercial profit. They are so scarce, indeed, that it is still considered quite remarkable for a man to own his own plane and to use it either for business or pleasure. This is particularly unexpected in view of the fact that flying is extraordinarily good sport and apparently quite appropriate to the pace of modern living. It is said to be a mechanical age, so that people should take quite naturally to the use of the flying machine for personal business and pleasure. But obviously they have not done so.

Nor is this because of the high cost of airplanes. Hundreds of thousands of Americans, for example, can afford two or three expensive automobiles, a garage to house them and a man in uniform to drive them. But a mere handful of them own an airplane and maintain a hangar and pilot. Many of those who use the airways quite readily for travelling show no apparent desire to purchase their own plane.

This is a curious outcome of a curious history, whose principal incident was the Great War and the effort to commercialize its experience and surplus materials in times of peace. The strictly commercial phase of flying is far ahead of private ownership and operation. And this, it seems, is one of the outstanding weaknesses of aviation today.

A fair comparison can be made between the automobile and the airplane and between their respective histories. Both are relatively small vehicles, compared with the railroad train and the ocean liner, and either can be operated by one man. The transport plane compares in utility with the big bus; the smaller plane has approximately the same passenger capacity as the private automobile. They use the same fuel; they have similar advantages over transportation that must follow the routes of rails or wires. They carry their own power and can go, more or less, wherever they please.

But the history of their development has been fundamentally different. The automobile was developed and improved in its early days by private ownership and operation. It had little commercial importance until it had served time as the "pleasure car," though the pleasure of operating it in its early days was sometimes doubtful. Study of the license records in the United States show very clearly that the privately owned and operated car kept the factories busy long before the commercial truck or bus was taken very seriously. Not until 1916, indeed, did the commercial vehicle begin to multiply rapidly on the roads, by which time the automobile was already in universal use as a private means of transportation.

The private car led the way in another respect. It demanded decent roads and eventually got them. When they were available everywhere, the truck and bus began to run on them. Now there is a serious danger that the increase of commercial traffic will crowd the so-called "pleasure car" off the roads, or at least make its operation hazardous and uncomfortable.

But airplane history reverses this story. A great effort has been made in many countries and for many years to make a success of commercial air traffic. Elaborate airways have been built to encourage it and every imaginable stimulus has been applied to it. But the privately owned and operated airplane is still a rarity and its visits are only incidental to the business of airports and the traffic of airways.

Very likely this is the cause of some of aviation's most obvious difficulties and the doubts of those who distrust its future. If there were tens of thousands of airplanes flying in the United States, instead of the five or six thousand actually in operation today, there would be no reason to question the policy of investing millions of dollars in the maintenance of Federal airways, with their elaborate and expensive systems of radio control, weather forecasting and night lighting. If there were one-hundredth as many airplanes produced as there were automobiles in recent years, the factories would have plenty to do. In the United States alone this would have meant an output of more than 50,000 planes in 1929. Actually in that period were produced between 5000 and 6000, a total that might be built in a single modern factory under the pressure of production methods. It is obvious, moreover, that a wide use of private planes would support many accessory industries, would provide employment for an army of skilled workers, and would justify an airport in every town and city. These, in turn, would inevitably reduce production and operation costs, so that the ownership and operation of a plane would be brought within range of the average pocketbook.

But all this depends on an immense increase in private ownership. Why has this increase been so long and stubbornly delayed?

The Autogiro suggests the answer and offers a remedy. It has been shown that among its most striking characteristics are ease of operation and inherent security for the man who flies it. The lack of these things has delayed more than any other factor the multiplication of the private owner and pilot. It is easily understood, for instance, that a man might feel able to afford to buy a plane but not be willing to spend the time and money necessary to become a thoroughly competent pilot. It is apparent that a man might trust himself to travelling in a transport plane with a highly-trained pilot and be quite unwilling to trust his safety of life and limb to his own amateur abilities in operating an airplane.

In these two respects, comparison between the two types of aircraft is striking and even startling. It takes not less than fifty hours of training to make a reasonably competent airplane pilot. A student may solo in a few hours, but this does not make him a pilot. In order to qualify for any kind of commercial responsibility, indeed, he will need not fifty but at least two hundred hours of air experience. And only a pilot is competent to drive an airplane, even though he is risking nothing but his own neck.

But the Autogiro can be operated with reasonable success and security by a man who makes no profession of being a fully-trained pilot. He can learn its essential operations in two or three hours and then be trusted with it in solo flights. A few hours further practice will give him mastery of the Autogiro under all ordinary conditions of flight and landing. In ten hours, at a conservative estimate, he should be able to fly it with confidence and a considerable degree of pleasure. He may not be skilful, but he will be safe. He may not fly very well, but he will fly without undue danger to himself or his machine.

The second major advantage is that of inherent security. It cannot be said too emphatically that the airplane is never safe in itself; it is safe only in the hands of a thoroughly competent and consistently careful operator. If a first-class pilot makes a mistake at the wrong time, there is likely to be a serious accident. Bad landing conditions, motor failure, unfriendly weather or any one of a number of possible errors of judgment may bring disaster to the best of pilots in the best of planes. This fact, which is substantiated by so many air accidents, is a profound discouragement to private ownership and operation. The average man will not trust his own abilities and experience in a machine whose safety depends so much on driver, skill, judgment and watchfulness.

But the Autogiro does not ask such extraordinary ability of its operator in order to keep out of trouble. An amateur pilot might get into trouble while flying it, but he can easily get out again. If his wits fail him altogether in emergency, he can come down with assurance that the machine will stay right side up and land without serious damage. Moreover he has a thousand chances of a fair landing where the airplane would look for one; any small field, lawn or open space will be room enough for the Autogiro to make a safe descent, though it may not be large enough for it to get out again. In flying near the ground, it needs only a small degree of skill to stop the Autogiro short of an obstruction, turn away from it or land in front of it, none of which are possible in an airplane under emergency conditions. A bad surface for landing is not nearly so dangerous to the Autogiro as to the plane. And more important, perhaps, in the aeronautical sense, is the fact that the amateur pilot of the Autogiro need have little fear of the minimum flying speed which is the worst hazard of the student airplane pilot. He need not be constantly aware of the danger of stall or spin. These deadly performances are inherently characteristic of the airplane, but the Autogiro is inherently averse to them and even incapable of them.

These are reasons why a man might acquire an Autogiro and be confident that he will be able to fly it soon and with a fair degree of security. The rest depends on practice. But the Autogiro owner will be using his machine for business and pleasure while he is practising toward perfection, just as an automobile owner starts using his car long before he is a first-class driver.

One further fact encourages the idea that the Autogiro is the family flying machine of the near future. Its ability to land in a small space makes it highly adaptable to private flying. Its owner would not need to route his travels by way of established airports nor to deny himself the use of his machine because there are no surfaced runways in the neighbourhood of his destination. An open field beside his mountain home might be his private airport; a lawn on a country estate would serve the same purpose. The seashore beach or a suburban meadow would offer safe landings for an Autogiro. The importance of this feature may be reckoned by imagining how automobile traffic would be restricted if it had no more freedom than the railroad train. The private car is useful because it needs neither tracks nor terminals. To a considerable extent the Autogiro has the same freedom, in comparison with the plane, which must land on an unobstructed space of many acres, with runways a thousand feet in length by which to take the air again.

The family flying machine must be comparatively easy to operate, inherently safe and reliable, and adaptable to a wide variety of landing conditions. The Autogiro is all of these. It seems, therefore, to offer a real prospect of widespread private ownership and operation, multiplying many times over the number of aircraft in use, encouraging the public use of aerial transportation of all kinds, and immensely stimulating the aviation industry in all its branches. It offers the immediate prospect of a real renaissance for aviation, which may lead to an "air age" that is worthy of the name. It promises a volume of air traffic that will make all the money spent on airways and airports seem a profitable investment. And it invites the general use of the air for business and pleasure, for travel and sport, for pleasure jaunts as well as for speedy communications.

In the fall of 1930 I left England in a European Autogiro and flew over land and sea to my home in Spain. Thence I took off on a tour of Spanish towns and cities, flying daily for more than a month and visiting about forty different places. I stopped at airports or in open fields and everywhere gave demonstrations and carried passengers. It was an entirely pleasant trip, made without elaborate preparations and without a single discouraging incident, almost as casually as a man might take a tour in a comfortable car. It was, indeed, an aerial holiday as well as a business trip.

When private owners, sportsmen and business men are doing the same sort of thing, flying their Autogiros for the fun and profit of it, the family flying machine will become the backbone of a great industry and also an invaluable utility and convenience of progressive civilization. Not everybody will fly, but flying will no longer be confined to commercial carriers and a few wealthy sportsmen. And very likely there will be ten private Autogiros in service for every commercial craft operating for hire, which is about the present proportion between the totals of privately owned automobiles and commercial vehicles on the highways.

XIV

AUTOGIROS AND THE AIR MAIL

IN some respects the air mail system of the United States is the outstanding success of commercial aviation to date. It has developed in a few years until its daily service covers over 85,000 miles, more than half the entire scheduled air transport of the United States. In the United States, the air mail service has been the prop and stay of general commercial traffic by air and of the aeronautical industry. The government, indeed, has frankly encouraged and supported the air mail for the general good of aviation. It is probable that this form of aid has proved far more satisfactory than the direct subsidies of Europe, particularly because it has allowed plenty of opportunity for business initiative and progress toward efficiency, which was rewarded with actual profits. It is certain that without the stimulus of the air mail, aviation in America would not lead the world today in airway mileage and traffic.

Development of the air mail service still goes on, with changes and improvements according to the progress of aviation and the experience of the operators and postal authorities. Enthusiasts for air mail are hopeful that the bulk of first-class mail matter may eventually be carried by air, for speed is a major consideration in such a service and the flying machine is capable of greater speeds over considerable distances than any other means of transportation. But at present it is not finally determined whether this is economically possible. It depends to a considerable extent on an increase in public patronage, paralleled by a decrease in the cost of carrying mail by air and distributing it to its ultimate destination.

At some time in the comparatively near future the American people will probably need to decide whether the air mail is to be considered only a luxury service for special sorts of mail, or is to carry the bulk of first-class mail and postal express. Economical operation depends very largely on volume of traffic, and it may prove impossible to reduce very much the cost of operating air mail services without turning over to them a much greater volume of business.

No mail system can be efficient unless it is comprehensive. It must reach to all parts of the country and to small as well as large communities. At present the air mail map of America, extensive as it is, is made up of half a dozen trunk lines and a few feeders. Only a fraction of the cities of the United States have direct air mail connections; many States are entirely aside from the air mail routes. Smaller communities can connect with them only by ordinary mail service to some central point.

If the air mail is ever to carry the bulk of first-class mail matter, the system must be enormously extended until it literally covers the country. This can hardly be considered possible under the present requirements of commercial aviation. Airways suitable for traffic by fast plane under a wide variety of weather conditions are expensive to build and maintain, but the air mail at present cannot travel safely and on schedule without them. And any other sort of mail service is a nuisance rather than a utility to the public.

The Autogiro may be the answer to an obvious problem in the extension of the air mail map. It can fly wherever there is air and an occasional patch of open field, and fly under these conditions more safely than the plane can follow an elaborate airway. It can land nearly anywhere without need of a large and fully equipped airport. This means that the smallest town can become an air mail station without overloading itself with the expenses of a municipal air field. It may be supposed that a dozen small cities, towns or even villages might be linked together with air mail service by means of Autogiros, with practically no expense to the communities themselves and very little to the government. In those parts of the country that have limited railroad service this would mean a major improvement to the mail service.

A further possibility is the development of devices for the delivery of mail without a landing and for picking it up in flight. Devices for air mail pick-up have already been developed, but no great use has been made of them. The problem of loading and unloading an airplane in flight is not easy, for the plane is moving at express train speed at all times. Moreover it is at best a matter of some difficulty and danger to fly close to the ground and make accurate contact with a loading device, taking up at the same time the sudden shock of an extra load. But I see no reason why this should not be much simplified by means of the Autogiro. Certainly a pilot could drop a mail bag from a hovering Autogiro with much greater accuracy than from a fast-moving airplane, and could do so at very little distance from the ground in perfect safety.

On the other hand, the slow-moving Autogiro could be brought easily and safely within reach of a loading device to pick up a mail bag without landing. On small feeder lines this might be desirable and would certainly reduce the operating costs of such a service and speed up the mail schedules. This is already done with ordinary mail travelling by railroad. The express train does not stop at the way station to pick up mail, but takes it automatically at full speed from a loading device. Without some such arrangement many small communities and rural sections would have a pretty poor mail service.

I believe that by the use of Autogiros the air mail service could be made much more comprehensive than it is likely to be for a long time if the mail is to be carried by planes. It may be that most of the first-class mail matter will eventually be carried by air, but not under the present conditions prescribed by the limitations of the airplane. And it is probable that a great increase in air mail volume would in turn reduce the transportation and handling costs.

There are good reasons for thinking that Autogiros will in time be generally adopted for air mail purposes on present lines, particularly where the route lies through difficult flying country. The essentials of any mail service are speed, reliability and security for the mail bags and their contents. Transportation of mail by air has given an advantage of speed, but regularity of schedules and deliveries has varied a great deal, principally according to weather conditions. The United States Postmaster General said early in 1931 that on some lines during the winter months mail schedules were as low as 15 per cent trustworthy. From this low figure they have varied all the way to approximate perfection.

Some failures to meet and maintain schedules have been the fault of poor co-ordination with ground transport of the mail. But a majority of them have been the result of bad flying weather, which has made it the part of common sense to keep the mail planes in their hangars. Fog and low ceiling often make it impossible for high speed planes to fly in reasonable security, and though the mail pilots are usually willing and anxious to go ahead, the more conservative operators will not permit them to risk their lives, their planes and their valuable cargoes in bucking bad weather.

Under some such conditions an Autogiro could go safely on with the mail, flying at low levels and low speeds which would be dangerous or impossible to the plane. An Autogiro would also have an important advantage in avoiding the dangers and delays of an unexpected storm or layer of fog encountered en route. Many mail planes have been reported missing after leaving an airport in fair weather; later it has been discovered that they

LANDING IN CLOSE QUARTERS

The Autogiro makes a successful landing immediately in front of an obstruction.

A TYPICAL TAKE-OFF

The Autogiro leaving the White House lawn after presentation of the Collier Trophy.

have been forced down or compelled to detour far from their course by a sudden change in weather conditions. Many mail planes have been lost for the same reason. Finding himself suddenly in heavy fog, the mail pilot must decide whether to take a chance on pushing through or immediately turn back into clear weather. An error in judgment may mean that he will be literally lost in the sky, and it must be remembered that it is extraordinarily easy to get hopelessly confused while flying in thick weather, and that it is often impossible to get out of it or make a landing with a reasonable chance of safety.

In some cases the mail pilots have had no choice but to leave their ship in mid-air, take a chance on a parachute landing and salvage the mail if they can. Others have fought their way through; some have crashed while trying to do so. In a number of cases the mail plane has been driven far off course by bad weather, so that the mail has been badly delayed and the schedule upset all along the line.

For some of these conditions there is no cure. There are some sorts of weather that prohibit flying absolutely, just as there are snowstorms that stop train traffic and fogs that tie up shipping. But there are many conditions of fog and low ceiling in which the Autogiro could carry on, flying slowly if necessary and keeping close to the earth where the air is likely to be clearer and where the pilot can hold his course by means of familiar landmarks. And if conditions get too bad, the pilot can put the Autogiro down nearly anywhere without danger to himself, his ship or his cargo. He can do so near to a highway or railroad, so that the mail can be quickly transferred to surface transportation to avoid long delays. And very rarely, even on a scheduled service through varying weather conditions and over difficult country, would it be necessary for a mail pilot flying an Autogiro to "bail out" at the expense of his craft and the risk of his cargo. Even if he did, the probabilities are that the Autogiro would land itself without any serious damage.

But even this possibility is a little fantastic. If the weather is fit for any flying whatever, it is good enough for an Autogiro landing. The country beneath the air mail route may be bad, but it can scarcely be so bad that a pilot could not find any field at all large enough to hold an Autogiro. Except over a great city or some stretch of forest, the Autogiro pilot has the choice of a thousand emergency landing fields. And if his engine keeps running—as modern aircraft engines usually do—he can pick and choose among them as he flies along at no faster pace than that of an automobile on a foggy day.

This comparison will serve very well to illustrate the advantage of the Autogiro in negotiating unfavourable weather. The driver of a car is not afraid of bad weather, though he may not like it, because he can control his speed and stop when he chooses. If he could not drive except at sixty miles an hour or more, he would not drive at all in fog and rain and storm. That is the plight of the airplane pilot, which is why the air mail is likely to stop running in bad weather or be seriously delayed by planes driven out of their course or forced down in emergency landings.

The Autogiro is ideally suited to the air mail service, particularly in those parts of the country where the weather is fickle and the countryside crowded with obstructions to ordinary airplane landings. This, as it happens, is the general condition in the busiest sections of the United States, where the most people live and where the air mail load is likely to be the heaviest.

An outstanding opportunity for the Autogiro to improve the efficiency of the air mail service lies in its ability to make landings near the heart of the modern city, where the greatest volume of mail must be collected and unloaded. Serious delays are involved in transferring the mail from present airports to central post-offices, which would be avoided almost entirely by the use of Autogiros landing down town on suitable fields or the tops of buildings. It seems logical, indeed, to expect that at some future time the roof of a post-office in a great city will be laid out as a landing field, not large enough for a safe landing for the plane of today but amply adequate to an Autogiro landing.

By the use of Autogiros for carrying the air mail, I believe the service could be easily and economically extended to cover a tremendous territory that is at present only indirectly in reach of it. At the same time the substitution of Autogiros for mail planes would raise the efficiency of the air mail by improving the regularity of schedules under adverse weather conditions. The air mail service depends for success on efficiency and reliability. In both respects the Autogiro promises to make a substantial contribution to an important service and public convenience.

XV

FIVE-ACRE AIRPORTS

TRANSPORTATION of any commercial type must have its terminals. An enlightening history of rail travel could be written around the development of great key terminals, which are like so many nerve centres to the networks of steel that cover the continents. After more than a century of railroad transportation, the building of bigger and more efficient terminals continues to be the most dramatic phase of its construction program.

Ports are equally important to transportation by sea. A natural harbour may be the cause of a city's greatness; the building of a breakwater and waterfront facilities may turn the tides of commerce in a new direction. Ships may sail anywhere on the sea, but they must bring their cargoes into well-appointed ports, the essential terminals of sea traffic.

Other familiar modes of transportation demand the same service, each according to its special conditions. The rise of bus traffic has forced a new responsibility on the modern city, which must make space somewhere near its centre for convenient passenger stations and union terminals. Even the private automobile needs parking space and the tendency of today is to require that it shall be stored at one end of its journey in a prescribed and protected area, similar in some respects to the terminals of commercial traffic.

Air transportation until now has been in particular need of a highly developed system of terminals. It has needed them so seriously, indeed, that they have been built already at enormous expense, far ahead of the air traffic that is expected to use them. There are, for example, several thousand airports and listed landing fields in the United States. Probably in 1930 there was an average of one such field for every three or four planes engaged in commercial air traffic.

This, of course, was only a temporary disproportion between air traffic and air terminals and way stations. But it serves to indicate that airplanes do not fly independently of conditions on the earth below. Even the cruising sportsman of the air, flying with no special destination or regard for schedules, keeps a wary eye on landing conditions beneath him and prefers to follow an established airway over unfamiliar or difficult territory. An airway is actually built on the ground. It is a linked series of airports and landing fields, often equipped with beacons for night flying and radio and weather reporting facilities. It is a highway of the air, but it is determined rather definitely by the "lay of the land" and by conveniences and safeguards like those which aid traffic by road and rail. And its terminals must usually be located in or near great centres of population, for the concentration of all transportation is to the cities.

This necessity has placed a big burden on the business of aviation. It has made it essential to find airport space in places which are already crowded close with people, buildings and business interests. Practical aviation has needed the downtown airport, but has been indifferently successful in getting it. The alternative is the semi-suburban airport or a field far out from the city centre, connected with it by high speed surface transportation. But these represent a compromise with the ideal. If commercial aviation could have all it wants, the transport, express and mail planes would land as closely as possible to the heart of the cities, where the activities of civilization are at their peak.

As a matter of present fact, the great airports of the world are not at the centres of the cities. London's airport is almost an hour's run from mid-city; Le Bourget is a long way from Paris; New York has no metropolitan airport, so that air traffic to the world's most crowded city must land about forty-five minutes away. Chicago's great field is an hour away from the "loop"; Los Angeles' field is "somewhere in the suburbs"; Philadelphia patronizes at present a field beyond the Delaware River and in another State. In nearly every large city of the world, the airport is at a considerable distance from the natural centre of traffic.

There is an obvious inconvenience in this condition, which has hindered the free development of commercial transportation by air. A traveller who is sufficiently in a hurry to travel by plane is likely to grumble at his slow progress when he disembarks at the airport and has still far to go. It is equally unfortunate that airports cannot be brought closer to other terminals of transportation. Railroads can come by tunnel to the very heart of a city, as is the case in New York City. But the Pennsylvania Terminal at 32nd Street, one of the world's biggest and busiest railroad stations, is at least forty minutes by surface transportation from the city's most convenient airport at Newark, N. J.

An ideal co-ordination of vehicular traffic is proposed by the plan for Philadelphia's municipal airport at Hog Island. When and if it is completed according to the program, air lines will meet there with transportation by road, rail and river. Passengers will be able to transfer from aircraft to fast trains, to smooth highways or even to ocean liners. Air express will go from plane to boat or train with a single handling. That, no doubt, is what is desirable for complete commercial efficiency. But the Philadelphia plan is far from realization and must cost at least $10,000,000 to carry out according to the complete specifications.

This illustrates the extraordinary difficulty of providing a downtown airport for a modern city, or one that can provide equivalent efficiency and convenience. Central city real estate is too high-priced for airport purposes, on the scale demanded by the airplane. Minimum requirements would mean the clearing of several city squares, merely to provide long runways in several directions. Nor is the runway the only unit of measurement for the modern airport. A space of a thousand feet square in mid-city might have room enough for runways, but not enough for clearance of the average city's sky-line, which groups its highest cliffs and towers at its main cross-roads.

It is beyond reasonable probability that the city of today will be able to find room for central airports. A new city might be built around a great air field, and ambitious young communities can plan now to leave room in their growth for an air terminal. But many millions of dollars cannot buy enough space for the purpose in the metropolis of today. The price is prohibitive, no matter what the scale of air transportation may be.

This problem has not been faced with entire honesty in the commercial air program of most countries. Efforts have been made everywhere to make municipalities or governments assume the burden of providing air terminals, since commercial aviation cannot afford them. Often they have done so, at considerable cost to the taxpayers; sometimes at the expense of future generations. Millions of dollars have been tied up in some airports which can never be repaid out of airport earnings. The only financially successful airports in America, indeed, are those that have attracted to their neighbourhood a number of aviation industries and amusement activities, whose ground rentals and leases or taxable valuations have provided some reasonable return on the investment.

The indirect advantages of air communications may justify municipal ownership and operation of airports. But the most air-minded city in the world would hesitate to tear down many acres of houses and business buildings and set aside its most valuable land in order that airplanes may land in the middle of things. Therefore the semi-suburban air terminal seems inevitable.

It is inevitable, however, principally because of the present limitations of aircraft. Among them is the airplane's need of long runways for the take-off and plenty of space for landing. It also demands ample clearance for a fairly flat glide into the field. High buildings, overhead wires and other obstructions around a flying field cut down its useful space considerably. Another factor to be considered is the air disturbance above the large city. It would be conservative to say, indeed, that an airplane needs two or three times as much clearance for safe operation as it actually needs for landing and take-off.

The present requirements for a first-class airport are formidable. Three or four hundred acres are not too much; a thousand would be better. Long runways must be available in several directions so that planes may land and take off head-on to whatever winds are blowing. There must be extra space in proportion to the traffic, for the chance of collision is to be reckoned with when several planes are coming in and out at once. The modern airport with any considerable traffic must flag its planes in and out, indeed, with as much care as the trains are signalled and routed in a rail terminal. The reason, once again, is a matter of airplane characteristics. Landing and leaving the ground, as they do, at anything from fifty to eighty miles an hour, they must be protected by plenty of free air around them. The pilot, moreover, has plenty to do and think about at the time of landing and take-off, without being crowded by other traffic.

We turn now to consideration of how the Autogiro will affect these conditions. If it will make possible and practical the downtown airport it will have made a substantial contribution to the efficiency of air transportation.

Personally I do not suppose that the Autogiro, even when it is in universal use, will justify large-scale airports at city centers. It is difficult enough to find space there for parking automobiles, and the ideal airport is not only a landing field but also a parking place and a service station for aircraft. But I foresee the early development of downtown landing fields for the Autogiro, on a scale much smaller than that of the true air terminal or airport. Since it lands nearly vertically and takes off at a steep angle after a ground-run of a few feet, the

Autogiro can get in and out of much less space than the airplane needs. It is not nearly so much cramped by surrounding obstructions nor by the erratic air conditions caused by them. It could therefore be landed with consistent ease and security in a fraction of the space needed for landing an airplane in mid-city—possibly in one-tenth the space.

Half a square mile would not be too much for an airplane airport in the middle of a city. Thirty acres would be plenty for an Autogiro field. This would allow a full thousand feet in any direction, and as a matter of fact an Autogiro can be landed easily in a two hundred foot circle under nearly any conditions, and can be flown out of any field that has five hundred feet of open runway. But for ample safety we may assume an Autogiro field of thirty acres. This is less than a tenth of what is required for an A-1 airport and much less than would be necessary for safe landing and take-offs for airplanes in the centre of a city.

I look forward to a new type of air traffic, whose destination will be the heart of the modern city. Room will be made there for Autogiros to land and discharge their passengers and packages, providing quick connection with other means of transportation and directly serving the convenience of the public. I think there will still be great airports around the cities or in their less congested sections, surrounded by all sorts of service facilities and spaces for parking and storage of aircraft. They will probably continue to be the chief terminals for long distance air lines, which will be supplemented by Autogiro service to the centre of the cities.

Private owners and pilots will, under such conditions, be using Autogiros for commuting, somewhat as the man of affairs makes use of his automobile today. They will fly down to the Autogiro airport, and their machines will be flown away to the hangars or storage spaces provided elsewhere, until needed for the return trip.

A prospective variation on the idea of downtown airports is the landing platform built above offices, railroad terminals and warehouses. These have been discussed rather more optimistically than is justified by experience with the airplane and the actual character of the modern city. Current city planning is not favourable to the flat roof, but prefers the architecture of pyramids and towers which

A SPORTSMAN'S FLYING MACHINE

The Autogiro flies along the beach at Miami.

FRANK HAWKS FLIES THE AUTOGIRO

Hawks says "The smoothest, safest landing of any aircraft I have ever flown."

gives light and air to the maximum degree, both for the tenants of the buildings and for the streets below. A landing field for airplanes on top of a city would very likely turn out to be an addition but no improvement to the modern metropolis. It would need to be of enormous extent, and planes would need to be much improved before they could be landed and flown away in consistent safety from such platforms.

But it is conceivable that a few city blocks might be covered in to provide a good landing for the Autogiro. If this were done, the business man might make a direct journey from his private flying field to the roof of his office. This is an entirely practical prospect. Whether it will be considered desirable by those who plan the cities of the near future remains to be seen.

At the other end of an Autogiro's morning journey from home to office there will be needed another sort of terminal. This will be, perhaps, a five-acre airport, a clear space with a suitable surface, unhindered by nearby obstructions and within ready reach of those who would use it. Thousands of private estates have plenty of room on their grounds for an Autogiro to land or fly away. In rural districts a field without trees or fences would serve the purpose. The little space needed for a landing field in open country may be illustrated by example from the first factory to manufacture Autogiros in America. Pitcairn Aircraft factory is located on a flying field of several hundred acres, laid out originally for the activities of a commercial flying service for airplanes. But only a corner of it is needed and reserved as an Autogiro landing field. No more is needed, provided there is free space in the neighbourhood for emergencies or for a miscalculated landing.

Between the private landing field and the city air terminal there are many variations of the airport, suited to every community condition. Every village and town could spare space in its neighbourhood for an Autogiro landing field, without incurring the expenses of an airport on the scale necessary for the airplane. As this is done, the air map will immensely outgrow its present proportions. It will not be made of a few trunk lines and their feeders, but every enterprising community will have a place on it. And when this comes to pass, aircraft will be able to use the air far more freely, sure of safe landing wherever it is needed and a port at the end of every aerial journey.

It has been discovered that the Autogiro excels the airplane because of the range of its efficiency in many respects. This range is equally apparent in its relation to airports. It can fly safely in and out of a five-acre airport, a private estate, a field in the open country. It can land in emergency in a space not much larger than itself. It can bring air traffic to the centres of population and industry and complete the co-ordination of modern systems of transportation. It is the answer to the high cost of air terminals, which has threatened to handicap seriously the commercial development of air traffic. It is the answer to the problem of the private owner and pilot, who might be able to afford a flying machine but not a hundred-acre airport from which to operate it.

XVI

SPEED, CEILING AND PAY LOAD

WHEN the Autogiro first attracted public attention there was general doubt, both within and without the aviation industry, whether it would ever amount to more than an interesting aeronautical experiment. Scepticism of this sort is natural and to be expected; it is also wholesome and for the general good of aeronautical science. The Autogiro survived it and rapidly won converts to its principle and the possibilities inherent in it. But some of the scepticism persisted when the facts of experience and demonstration had proved certain criticisms to be unfounded.

It was often said in the early days that the Autogiro could not fly fast enough to become an acceptable substitute for the airplane, no matter what other characteristics it might possess. This criticism was based on the fact that the earliest experiments attacked the principal problem of flight by means of rotating wings, and made no effort to attain any impressive speed. It was first important to make the machine fly and fly properly; later it could be developed to fly fast.

In 1925, for example, the top speed of the Autogiro was about seventy miles an hour. In 1931 an Autogiro can fly at 120 miles an hour, nearly double the pace of its early predecessor.

What are the prospects for further development in this respect? It may be confidently anticipated that the Autogiro will eventually catch up with the airplane and under certain conditions exceed its speed performance. This probability is implied first by the many possibilities of streamlining the Autogiro, a process which has repeatedly improved the pace of the airplane without any increase in power. Quite a few obvious opportunities for such refinement are apparent in present models. The present under-carriage, for example, induces a considerable drag. Whenever speed becomes of sufficient importance to justify the study involved in this problem, it may be expected that this will be modified accordingly, as the fastest planes of today have added not a little to their speed by simplifying their undergear or providing for its retraction into the body of the fuselage in flight.

A measurable amount of drag is induced by the cables which at present support the rotor blades while at rest and prevent them drooping to the ground. The same is true of the friction dampers which regulate the lateral movement of the blades. Experiments under way in England in early 1931 hold promise of eliminating these features, a process which may be compared to the gradual elimination from the airplane of wires and struts in the wing structure.

Other possibilities of streamlining are suggested by prospective improvements in the present pylon which supports the rotor and by modification of the blades themselves. In other words, the Autogiro has by no means exhausted its possibilities of refinement for the sake of speed. But it may be fairly said that the fastest airplanes have nearly done so. The speediest craft of the current year are simplified to the point where they are superficially little more than a body as clean-cut as a bullet and a single wing. Further speed depends almost entirely on additional power.

In the higher speed ranges there is a theoretical probability of advantages for the Autogiro which have no equivalent in the plane. A comparatively simple improvement in performance is involved in the slight flattening of the disc of the rotor at higher speeds. More important is the diminishing effect of drag. The drag or effect of air resistance increases according to the square of the speed, which is one reason why improvement in this respect slows up progressively until no further speed is practical in terms of efficiency. Comparing the Autogiro and the airplane, the drag induced by their fuselages and control surfaces increases equally as greater speeds are attained. But this is not true of the comparative drag on the wings of the plane and the rotor of the Autogiro.

The amount of drag depends on the actual motion of the airfoil and increases in proportion to its increased speed. When an airplane flying at 100 miles an hour increases its speed to 110 miles an hour, its wing speed has been increased by 10 per cent. But the speed of the rotor blade is a composite of the forward speed of the craft and its own rotary motion. If the blade tip is moving around its axis at 200 miles an hour and the craft at 100 miles an hour, the forward speed of the blade at one point in its cycle is 300 miles an hour. An increase of speed of the whole craft of 10 miles an hour, therefore, increases the blade's speed by only about 3 1/3 per cent, and the increase of drag is in the same lesser proportion.

This is important because it implies that the efficiency of the Autogiro will measurably increase in comparison with the plane when high speeds are attained, whether by improved design or additional power. From present knowledge and experience curves have been plotted which demonstrate this probability. The drag-lift coefficient of a plane at a steadily increasing speed may be represented by a continuous line. At low air speeds, of course, the Autogiro's curve shows a wide margin of superiority; at some speeds that are entirely possible to the Autogiro the airplane has, indeed, no useful lift at all. At the intermediate section of moderate speeds the airplane shows some superiority in respect to the proportion between lift and drag. But at greater speeds the Autogiro's curve approaches that of the airplane again, and may be presumed to pass it into a position of definite superiority when speeds in excess of 140 miles an hour are attained.

It has been questioned whether the Autogiro is capable of altitudes to be compared to those of the airplane. The point has not been proved by experience, but there is no reason why an Autogiro should not fly to the same height as the airplane, if designed and equipped for the purpose. Its present known ceiling is ample for all practical purposes of flight, and the problem of altitude flying as such has not been specifically studied.

The highest measured flight by Autogiro to date was made at Pitcairn Field by Miss Amelia Earhart in April, 1931. In a standard commercial Autogiro she reached a height of 18,400 feet, as confirmed by barographic records. In her opinion this was by no means the ultimate ceiling of the craft she was flying. It is interesting for the sake of comparison to recall that it was not until 1918, fifteen years after the invention of the airplane, that official altitude records were established beyond the 20,000-foot mark.

In respect to pay-load there is probably no important difference between the capacity of an Autogiro and a comparable airplane. There is, however, a marked superiority for the Autogiro if it be compared to airplanes designed to imitate in some degree its performance in the low speed ranges and in landing. The only planes capable of this sort of competition are light-loaded craft, usually unable to carry more than a pilot and a single passenger. A large-sized Autogiro with a heavy load flies as well as the transport plane and lands more slowly in less space than the specially-designed light-loaded planes that are actually little more than powered gliders.

Improvement in the load capacities of aircraft depend much less on design than on efficiency in power plants. At present the carrying capacity of the most efficient airplanes lies somewhere in the neighbourhood of one-half their own weight. If their weight is reduced without cutting down their power, this proportion can be much improved. Recent developments of Diesel engines, particularly in

Germany, seem to forecast a very substantial improvement, which the Autogiro will be able to share.

It is supposed that the useful load that can be carried by a plane is slightly but steadily bettered as planes are built larger and with greater power. Theory shows that this will be more true of the Autogiro. The indications are, in fact, that when bigger Autogiros are built they will also be better Autogiros, in respect to all the characteristics that make aircraft useful and efficient.

XVII

INDUSTRY AND THE AUTOGIRO

IT is generally recognized that the airplane's most important contribution to the service of civilization depends upon its speed. Because it is faster than any other means of transportation, it is superior to others for certain work and under certain conditions. It cannot carry the cargo of the railroad train, but it can beat the express train's schedule. Its useful load is about the same as that of the automobile, but it is much faster than the automobile and is in addition independent of highways, hills and traffic congestion.

The advantage of speed is indicated by the success of air transport in territory that is well served by other means of transportation. There is excellent railroad service out of Chicago in all directions, but many air lines radiate from that city and airplanes follow closely the routes of the express train services. Plenty of fast trains run between New York and Washington and the highways between the cities are excellent, but a busy schedule of transport planes is maintained over the same route. There is a proportion of the travelling public that welcomes every opportunity to cut a few hours from a journey, to avoid the necessity of travelling overnight or to arrive at a distant destination before the day is too far spent for business or pleasure. These are the real patrons of air transport, though there are many, of course, who choose the air route for its novelty and sometimes for its superior comfort. But the real travellers prefer the plane because it is faster than anything that runs on wheels or rails.

The speed of aircraft has proved equally useful to special needs of industry. Of this fact the air mail is an excellent illustration. The only advantage in carrying a letter by air is the time that is saved. But the airplane has also proved its worth in carrying such diverse commodities as medicines, machinery, bank clearances, advance fashions, motion picture films, news photographs, foods and flowers, and a great variety of express shipments in emergency or where time was an important consideration.

It has been shown that the Autogiro can compete successfully with the airplane in all the principal fields of flying activity. It can do so because it has a similar advantage of speed over surface transportation and has also many characteristics that the airplane does not possess.

But the advent of the Autogiro opens up many possibilities of usefulness that do not depend upon its ability to fly fast but are the result of its ability to fly slow. All the benefits conferred by speed are still available, but there has been added the great advantage of being able to move about in the air at a moderate speed, taking every advantage of the flying machine's mobility, independence of highways and ability to climb high above the earth, without the necessity of travelling always at more than a mile a minute.

Speculation might go far in discussing what the Autogiro will eventually mean to the world's business. But even in the early days of its commercial service there have been discovered many new possibilities of usefulness for aircraft to the complex activities of modern civilization.

One of the first to be recognized was the superiority of the Autogiro for aerial photography. Photographs from an airplane must be made with an extraordinarily short exposure and at a considerable distance from the subject, for the simple reason that the camera is necessarily moving more than a hundred feet per second. But the ability of the Autogiro to fly at a quarter of the airplane's speed and even to hover briefly in the air makes possible a new technique of aerial photography. An early demonstration of its possibilities was the making of direct color photographs from the air. Since color pictures must be taken through screens or ray filters, the light that reaches the plate is very much reduced in actinic strength, which means that the exposure must be lengthened accordingly. At the slow speed possible in the Autogiro relatively long exposures can be given, and by this means some fine color photographs were made from an Autogiro early in 1931.

Another recognition of the possibilities of the

Autogiro for picture-making was the fact that the first Autogiro produced for sale in the United States was purchased by a newspaper, the *Detroit News*. It was stated that this machine would be used to transport reporters to places difficult to reach quickly by ordinary transportation and also for obtaining unusual photographs of news events. Plenty of airplane pictures have been taken and published, of course, but a photographer working from an Autogiro can come to much closer range with his subject without taking chances on his own safety or on the results of his exposures.

A somewhat similar possibility was suggested by an explorer returning from Central America, where he had been surveying the jungles from the air for traces of the ancient Mayan civilization. A powerful airplane was used for the purpose, but the results fell short of expectations. Flying at sufficient height for safety above the tangled jungle, only the more extensive ruins of the Mayan cities could be located, and where others were suspected it was nearly impossible to make sure of their presence while flying low at high speed. The explorer, in reporting his experiences in an American magazine, expressed the opinion that the Autogiro would permit much more accurate observations because of its ability to fly slowly at the pilot's discretion, while at the same time providing all the advantages of the airplane in surveying inaccessible country.

The same advantage is offered in using the Autogiro for all sorts of surveys, in which the airplane has already proved serviceable. Airplanes are similarly used to patrol forest areas for the detection and control of fire, and here the Autogiro offers a special advantage because of its independence of large landing fields and the consequent increase of safety for the man and the machine over varied country.

It is probable that in the near future the Autogiro will be used for a purpose particularly useful to public utility companies in the United States. In this country there are great networks of wires and pipes which carry power and fuel from sources of supply to central points of distribution. There are also the pipe lines for transporting oil, some of them a thousand miles in length or more. Lately there has been a similar development of piping for natural gas, which is bringing the immense reserves of certain southern and western States directly to the great cities, where the gas is blended with artificial gas to supply an efficient and economical fuel.

Constant inspection is necessary along these trunk lines of wire and pipe. Over long stretches this is being done with considerable difficulty, usually on foot and over the most primitive country. But recently devices have been developed which are capable of locating breaks or leakage in power lines and even in fuel pipes by means of radio. It is quite probable, therefore, that an aerial patrol could be maintained by Autogiro which would not only result in great economies but would be immensely superior in speed to ordinary methods of inspection. This could scarcely be accomplished with the airplane. It must travel too fast and too high to follow closely the trail of a wire or pipe over the difficult country where these arteries of power and fuel are sometimes laid. But the Autogiro could follow these lines at low speed and near enough to check their condition both by direct observation and by means of instruments, under conditions which would be considered dangerous for airplane flight at low levels.

A practical use already proposed for the Autogiro in connection with the air mail is the efficient linking of mail steamers with the shore. It has already proved practical to send special mail ashore by plane, when an incoming liner is within several hours' flight of the land. But this has usually meant that the liner must carry a plane and pilot for the purpose and it has not proved practical to arrange two-way communications. Once or twice a pilot has attempted to follow a liner out to sea and drop a bag of mail on its deck, but usually he has missed his target and the mail sack had to be fished out of the sea.

Probably every mail-carrying steamship from the important ports of Europe has some mail aboard that demands the utmost haste, so that it would be worth while to find a short cut to avoid the delays that are likely to arise in entering harbour. Similarly, there is sometimes a proportion of mail that arrives too late for the liner and might be sent after it by air. The Autogiro may be the answer to the problems involved. Certainly it would be able to make a much better job of deliveries to a ship at sea and probably it would be unnecessary to halt the liner for the purpose. The Autogiro can fly safely at the normal speed of modern passenger steamships, so that it could be manœuvred directly above the deck and held there while the mail sack was delivered. Similarly a pick-up device could be devised for taking on mail from an incoming boat, making it unnecessary to carry a plane on board for express mail deliveries at the end of the journey.

The practical possibilities of the Autogiro for mail carrying were indirectly recognized recently in discussions of plans for the new post-office building in Philadelphia, Pa., and in other American cities, and are under active discussion in England. It is proposed to provide a clear roof level for landing air mail, on the assumption that aircraft will eventually be able to descend in mid-city to a comparatively small field. Nothing in present airplane development indicates any early likelihood of this being accomplished. But with the Autogiro it is both possible and practical.

Experiments are expected in the near future in

the landing of Autogiros directly on the deck of aircraft carriers and battleships. Naval aircraft can fly at present away from the deck of a warship, but must usually land on their return in the water and be lifted by cranes to the deck. A successful solution to this problem suggests a further variety of new uses for the Autogiro, depending on its ability to land at slow speed in a small area.

It is because the Autogiro has a so much wider range of useful characteristics than the airplane that it may be expected to multiply many times over the present services of aircraft to business and industry. Because it flies and because it is fast it offers all the advantages of the airplane. Because it can also fly slowly and land safely in a limited area it can be made useful in many ways that are intrinsically impossible to the airplane.

XVIII

THE FUTURE OF THE AUTOGIRO

WE have attempted to survey the history of the Autogiro from the beginning and its present uses as a practical flying machine. It may be interesting to speculate on its future.

There would be little purpose or profit in doing so in a merely imaginative fashion. Whatever is to be said, indeed, should be based on present knowledge and its indications. Some opinions may be offered as scientific probabilities, derived from the present performance of the Autogiro and the theory which accounts for it. Others have to do with the prospects of progress as they depend on human effort, intent on the final mastery of all the problems of mechanical flight.

I foresee no arbitrary limit to the size of the Autogiro, within the reasonable proportions allowed to heavier-than-air flying machines. In the near future we may expect the building of transport Autogiros, capable of carrying as many passengers as the bigger planes of today. Modifications will be necessary in rotor design in larger craft, but the principle of flight by autorotation can be applied to all sizes of Autogiros, from a small, high-powered and high-speed machine to the larger commercial carriers.

Cabin craft are, of course, entirely practicable, and one has already been built. Nothing now stands in the way of an Autogiro designed to carry passengers in the comfort and comparative quiet of an enclosed cabin.

The Autogiro principle can be efficiently applied to amphibian and watercraft. Autogiros with floats for alighting on the water have already been built in England, where they have been given the name of "hydrogiros." The difference between such craft and the ordinary Autogiro are likely to be no more than distinguish the airplane from the seaplane. It may be, however, that there will be a less pressing need for the development of the hydrogiro, because of the wide range of choice in landing possessed by the Autogiro. There are parts of the world where seaplanes or flying boats are to be preferred to airplanes, because of the presence of water and the lack of large landing fields. But in most of these there are spaces sufficient for an Autogiro landing, and since there is more land than water everywhere but on the ocean the craft that can make a landing on the solid earth has the widest range of usefulness.

The type of Autogiro presented today for commercial or private use is the type of greatest immediate usefulness and general advantage. Many other types are possible and will be developed as the need arises. And it is important in this connection to note that present knowledge is sufficient to guide the design of a great variety of Autogiros. As with airplanes, we can, with a fair degree of accuracy, predict the performance of various systems of rotors, calculate the speeds they will permit and adjust other details of Autogiro design accordingly. These predictions are the result of studies in the theory of the Autogiro, based on actual experience and constantly checked by results. They correspond to the elements in aeronautical science which direct the efforts of the designer of a new type of airplane.

But in spite of these certainties, I am well aware of the variety of possibilities still latent in Autogiro design. These are partly due to the fact that the Autogiro has many more parameters or variable elements than the airplane. In the rotor alone there are possibilities in the blades, their airfoil, length, weight and internal structure. The fixed wing and controls are equally susceptible to modification for the sake of improvement or special purposes. And all improvements that have been applied or may be applied to airplanes in respect to engines, instruments, fuselage design and structure and devices for control can be equally well adapted to the service of the Autogiro.

All these belong in the province of scientific probability or possibility. It may be that they will lead to types of Autogiro as different from the present craft as the modern airplane differs from its ancestor of twenty years ago. I sincerely doubt that aeronautical science will come to a standstill for a long time to come, and I believe that much of its advances will benefit the Autogiro.

But the other element—the possibilities inherent in human effort, ambition and competition—are equally important to the future of the Autogiro. And for this reason I believe it is profoundly important that the present policy of those in charge of its development encourages the co-operation of many minds in its perfection.

I may be permitted to recall the fact that in its early stages the Autogiro was a one-man product. It remained so until the essential principle of flight by autorotation was well developed and demonstrated. In recent years further experimental work has been carried on in several places and in several countries. A particularly intensive effort was made in America to perfect the craft in detail and prepare it for commercial production. But at the end of 1930 there were still only a handful of men in the world at work on the Autogiro, compared to the great army of designers, experimenters and inventors everywhere at work on the perfection of the airplane.

It seems certain that the policy which invites responsible manufacturers to continue investigation into further development for the Autogiro must result in rapid advances, some of which it is at present quite impossible to foresee. There is nothing like the pressure of competition to inspire originality and hasten the pace of progress. With many minds at work on it, I think we may be reasonably certain of both change and variety.

The Autogiro comes into the picture at a time when aviation has already found plenty of work to do. But it would be contradictory to the whole history and spirit of aviation to suppose that aeronautical science and business enterprise will be content with present accomplishments. The Autogiro, I believe, will come quickly into general use for nearly every purpose of flying that is now known. But many and varied possibilities lie ahead of it, hidden at present in tomorrow but hinted at by what is happening today.

I hope to have my share in this future. There will be very many others working in different directions and with different intentions and inspirations. It was by such collective effort, inspired by competition and guided by a great variety of experience, that the airplane came from its crude beginnings to the extraordinary accomplishments of recent years. It is my sincere hope that the Autogiro may profit as much by the genius and diligence of the designers of tomorrow.

The future of the Autogiro belongs now, indeed, to the public that will use it, the men of business who will develop its manifold services to civilization, and the designers and engineers who will further perfect its performance, structure and design.

XIX

THE WINGS OF TOMORROW

NO product of human ingenuity has ever been so tempting to the romantic imagination as the flying machine. Long before men could fly in fact they did so in fancy, filling their dreams of the future with winged wonders, soaring boldly and swiftly in the sky on the unknown business of unborn civilizations. Poets and philosophers shared such dreams with inventors and men of affairs. Tennyson talked of "heavens filled with commerce": Leonardo da Vinci and other masters of the arts and sciences gave some of their genius to devising flying machines. Jules Verne allowed imagination full rein when he told a tale of flying ships, and many imitated him for better or worse. Kipling wrote a thrilling and prophetic story of aircraft before the Wright brothers were in sight of success, and lived to learn that in many respects he was too conservative in his expectations.

The wish to fly, indeed, was father to the fact. Because men wanted wings, they found out at last how to make them and use them. But though the dream came true, the dreaming went on. The romancers have been busier than ever since the dawn of the Twentieth Century, and have spread strange fancies in the illustrated magazines, the newspapers and the pages of popular fiction. Not yet satisfied with the wings of today, they speculate freely and sometimes wildly on the wings of tomorrow.

A degree of damage has been done by such speculations, some of which have been inspired more by courage than conviction and more by optimism than understanding. It is not always wholesome to expect too much of the future, for this may lead to wasted effort in following some will-of-the-wisp into a quagmire of failure. There is also a more practical danger in excessive optimism. The aviation industry has suffered rather seriously in recent years from great expectations which have disregarded the sensible facts of science and business. Within recent memory, money has been heavily invested in all sorts of aeronautical absurdities. Some of it has been for research, more or less legitimate; more of it has been public money gambled too carelessly on the future of flight, principally because the public seems to grow more credulous as science performs new miracles.

Those who know least about aviation are likely to expect the most. They see what has been done already, and expect the pace of progress to continue and increase. Therefore they are likely to swallow whole the extraordinary prophecies of skilful journalists and pseudo-scientists who actually know very little more about aviation than they do themselves. They are likely to accept without question a fantastic picture of the future which ignores nearly everything that is certainly known about aviation.

We need a stiff dose of scepticism to avoid absurdity in speculating on the wings of tomorrow. We should not put too much faith, for instance, in the idea of super-planes as large as ocean liners until somebody has flown and operated at least one over-sized transport plane in unquestionable safety and at a business profit. We should be wary of tales of high speeds to come, noting that every other means of mechanical locomotion has soon found its speed limit, beyond which efficiency is impossible. We should doubt very decidedly the early prospect of the "fool proof" plane, for nothing remotely resembling it has been developed to date. Many of those who have claimed its invention, indeed, have eventually smashed themselves beyond repair in their own "fool proof" machines.

Nothing can properly be considered probable which is at present impossible. Nor will the facts of the future contradict the facts of the present. The fundamental laws of mechanics, physics and aeronautics do not change, and nothing is gained by imagining machines which are supposed to deny them. It is easy and exciting, for instance, to imagine a flying machine which is independent of gravitation and to contemplate what it might be able to do. It was done long ago and is still being done. But this is not forecasting the future—it is romantic fantasy, for we know by sufficient and invariable experience that the law of gravitation is

AVIATION PIONEER INSPECTS THE AUTOGIRO

Orville Wright and Harold F. Pitcairn with the first Autogiro brought to the U. S. Langley Field, Va., May, 1929.

JAMES G. RAY AND THOMAS A. EDISON

The distinguished inventor approves the principle and performance of the Autogiro at Newark Airport, 1930.

beyond our power to repeal or control, and likely to remain so.

So when the Sunday supplement of your favorite newspaper breaks forth with pipe dreams and pretty pictures of what the world may be like tomorrow—or the day after—it is the part of common sense to call it all a fairy tale and take it no more seriously than is necessary. Probably there are not ten sentences of sound aeronautics in all its prophecies.

Not long ago, a highly capable designer of stage scenery forecast in an American magaizne that within ten years "airplanes will be able to land and take off vertically." The gentleman makes a very common mistake. He supposes that because a thing is desirable, it will be done. What he omits to consider is that it cannot be done, under the terms of his proposition. His idea is admittedly attractive. But it is aeronautically unsound, for it is inherent in the nature and principle of the plane to fly like a powered kite and not to climb like an elevator.

Other men of less intelligence expect more startling performances, but make the same fundamental mistake. They ignore all science, except the ability of science to do remarkable things; particularly they ignore aeronautical science, which is an exact and exacting system of knowledge and experience. Yet even in our optimism we should aim to be intelligent and reasonable and leave room for certainties in our speculations, even though they concern so romantic a topic as the wings of tomorrow.

Other chapters in this discussion have outlined the limitations of the conventional airplane and discovered that they have not been in the least modified by a quarter century of progress. There are limitations inherent in all flight by means of fixed wings, which demand high speed for full flight and landing and involve uncertainty of control at less than a minimum pace. These deficiencies may be patched up, but they cannot be cured or avoided except by forsaking the fixed wing for something better.

But it has been discovered and demonstrated that the revolving wings of the Autogiro involve a new principle, contradicting none of the laws of flight but using them in a new way in a new type of flying machine. The Autogiro can do nearly everything that a plane can do. In addition it can fly slowly, descend vertically, land nearly anywhere and be flown by nearly everybody.

It is not too much to say that its revolving blades are the wings that the world has been waiting for. The prophets and romancers may amuse themselves still with fantastic forecasts, but the practical flying machine is a reality here and now. It is not yet perfected beyond reach of improvement, but it is ready for business. It may change as much or more than the airplane has changed since Kitty Hawk, but its principle is proved sound and will not be lost in the process of improvement. Its revolving blades may seem strange to the eyes of today, as the first airplane seemed crude and clumsy twenty years ago, but they are in all essentials the wings of tomorrow.

We may forecast, therefore, an immediate future in which aircraft will be of several sorts. There will very likely be great airships in the sky, cruising at comfortable speeds but at a very considerable cost per passenger and per pound of freight. They will link continents over the oceans, cross the polar wastes and the earth's rough places, and discover new possibilities of travel and expensive amusement.

Then there will be airplanes of extraordinary speed and mobility, designed and built for special purposes. Some will be units of fighting fleets of the air; some will be express messengers of the sky, as obviously designed for such service as the racing automobile or motor boat.

But the overwhelming majority of commercial and private aircraft will be Autogiros or something fundamentally like them. They will throng the air and come and go from ten thousand airports in every civilized country. They will be found with folded wings in the hangars of country homes; they will bring commuters from far away to cities of business and centres of industry. They will carry mail over thousands of miles of new airways. They will patrol the forests, the frontiers of nations, the long lines of utility services, the coasts and waterways. They will deliver far places from loneliness, and make near neighbours of the communities of a continent. They will fly with sufficient speed and security wherever there is clear sky above and a little space on the earth below. They will fly about the world's swift business on revolving wings —the wings of tomorrow.

Yet this is not to say that they will make obsolete any of the world's familiar modes of transportation. It is a familiar fallacy of those who are over-anxious for the arrival of the air age that the sky will soon be clouded with crowding aircraft, a tangled sky traffic of impatient speed presenting all sorts of problems for its control and direction. Common sense and experience say otherwise. No sound mode of travel has ever been put out of business by another. One type of transportation complements another and the efficiency of both is increased. The automobile, the bus, the railroad, the subway, the ship and the flying machine are all servants of civilization, and likely to remain so. Some are old in service, others are still seeking their proper place in a co-ordinated system of modern communications. One is just at the dawn of its development. This is aviation, which has worked and waited a score of years for the right wings.

Earth and ocean and sky will all be highways

for tomorrow's travel. Great hulls will plow the seven seas, wheels will roll on rails and roads, wings will soar everywhere in the freedom of the heavens. And the universal wings of tomorrow will be like those which turn today above the Autogiro—the product of science and experience and modern mechanical progress. They are safe, they are efficient, they are soundly scientific. They are the wings of tomorrow.

XX

THE THEORY OF THE AUTOGIRO *

SOME TECHNICAL CONSIDERATIONS OF FLIGHT BY AUTOROTATION

"AUTOGIRO" literally translated means "self-rotation." The syllables "giro" do not imply gyroscopic phenomena, but gyration. Self-rotation, applied to aerodynamic phenomena, is called "induced rotation" or "autogiration."

The Autogiro solved all the problems which defied the previous experimenters of ornithopters

* This chapter originally appeared in *Fortune* for March, 1931.

and helicopters in their endeavours to produce flapping and rotative motion. In this free system of rotative wings with freedom to flap, the blades actually do flap in the manner desired by the ornithopter advocates; at the same time the blades do rotate in the manner desired by the advocates of the helicopter. This is accomplished without most of the mechanical difficulties encountered in applying motive power because in this case the reactions of torque are eliminated and the blades are driven by the powerful air forces.

Among the array of phenomena which are constantly at work during the operation of the Autogiro, perhaps none is quite so difficult to grasp as that of the automatic rotation of the blades without a visible means of power.

An airfoil moving through the air is acted upon by air forces which have magnitude and direction. The individual blades of the Autogiro are so mounted that the lift or weight which is hanging on them, when revolving, results in a slightly forward slope or direction of this force, which is so powerful and so positive that the blades are obliged to move forward, and since their forward motion is defined by movement around the central axis, they have free rotation.

To obtain autogiration, it is essential that the blades should first be brought into a rotary motion by mechanical, aerodynamic, or other means until sufficient speed of rotation has been attained to effect some lift. While the machine is in flight, the forces on the blade can be considered as follows (see Figure I):

From the first condition of motion of the relative wind along BB, the combination of rotary velocity of the blade with either a forward or downward velocity of the machine as a whole gives the direction of flow as shown by CC. There is a lift reaction L perpendicular to this line, and a drag reaction D parallel to it. To obtain autogiration, it is essential that the lift reaction be tilted ahead of the axis of rotation a certain amount. This amount must be sufficient that, when a vectoral addition of L and D is made, the resultant reaction must still be ahead of the axis of rotation. As long as this resultant is ahead of the axis of rotation, the rotor as a whole will accelerate. When the speed of rotation becomes such that the resultant reaction is parallel with the axis of rotation, the system is in dynamic equilibrium and no further acceleration will result. With the resultant behind the axis, a deceleration or loss of speed in the rotor will result.

This combination of forces is identical for all conditions of flight, including take-off and climb, level forward flight and vertical descent with no forward velocity. The resolution of forces is the same except for some variations in their respective magnitude and direction. These variations are manifest in a final slight difference in rotational speeds only. The quantity of sustaining lift in the rotor itself remains virtually constant.

When the Autogiro descends vertically there is no dissymmetry of lift, all blades acting equally. But as soon as the device moves forward through the air at any speed, there is a velocity differential on the advancing and retreating blade, in which case the rotative speed is additive on one side and subtractive on the other. Since all lift is dependent on relative motion, from first principles, it is apparent that that blade having greater relative motion has a greater quantity of lift.

Hinged or articulated blades permit the yielding and rising on the advancing side and the descending on the retreating side, which, in effect, is flapping, and in this manner the blades automatically adjust themselves to the difference of air flow.

This is accomplished by allowing the advancing blade to rise as it moves forward, consequently working at less effective incidence at the faster resultant air speed. Conversely, the retreating blade descends as it moves toward the rear, the result being that it works at greater effective incidence in the less resultant speed. In this way, the advancing and retreating blades all carry their proper share of lift, and there is no force from the rotor tending to overturn the Autogiro at high forward speeds.

Concurrent with the solution of the dissymmetry of lift, this very hingeing of the blades eliminated entirely the problems encountered in gyroscopic phenomena which exist in every rigidly mounted rotating system. All gyroscopic action is eliminated through the attachment of the blades to the hub by means of articulating joints. The gyroscopic properties of a non-articulated rigid rotating mass as large and as heavy as the Autogiro rotor would be impossible to control. But when the rigid rotating mass is articulated, the gyroscopic precession and other undesirable phenomena from the mass are eliminated.

Such an undesirable system may be represented as a disc (Figure II). The thrust from the motor would cause a large gyroscopic force, F_g, to act. As a result, the machine would roll to the right. Moreover, the lift from a rotating rigid blade system in forward flight is displaced laterally from the axis of rotation. The tipping moment from the unsymmetrical lift will also generate a gyroscopic moment which will upset the machine longitudinally. A similar action would take place when any banking moment becomes introduced for control purposes.

In contrast with this is the hinged rotor blade system as represented in Figure III. When the machine is in forward flight under the action of motor thrust, each of the blades automatically assumes its natural position. Instead of the gyroscopic forces, aerodynamic and centrifugal forces are predominant, and these assist, in fact force the blades to follow the guidance of the axis upon which they rotate.

Next to unfailing sustentation, complete controllability and stability in the air are of paramount importance. In tribute to the high degree of perfection which this art has attained in the airplane, it will suffice to say that some modern-day airplanes can fly themselves for indefinite periods unattended by pilots. This type of stability, of course, does not remove the danger of the tailspin nor is it effective at the slower speeds.

In the Autogiro, a condition of stability is achieved which not alone rivals the perfection of the airplane at high speeds but exceeds it because the machine cannot stall dangerously or spin. At slow speed, it is under close control by the pilot. In vertical descent, its pendular stability is of the highest order, since the arrangement of the blades, their hinges, and their axes preclude pendulum oscillation such as is met in the parachute, while the distance between fuselage and rotor gives excellent leverage upon which the pendular weight can exert its stabilizing effect. For the first time, a flying machine can enter the air with a definitely unchangeable relationship of two points: one, the centre of gravity of the machine which represents all of its weight; the other, the metacentre of pressures which represents its lift.

To this condition of stability of the Autogiro, let us contrast that of the airplane, in which the centre of gravity remains fixed, but the centre of pressure moves not only fore and aft on the individual airfoils, but moves laterally across them, with various manœuvres calling for stabilizing forces or the attention of the pilot.

In Figure IV are shown rotor vectors in comparison with the centre of pressure locations of an equivalent fixed wing at practically equivalent speed conditions.

The rotor alone is essentially a stable aerodynamic surface. The reason for this is not a matter of centre of pressure travel fore and aft along the disc, as in an airfoil. Instead, as the forward speed of the disc is increased, the rotor thrust-line moves forward with an increased angular tilt, which also moves the thrust-line forward with respect to the centre of gravity of the Autogiro as a whole. As forward speed increases, the effect is a longitudinal moment from the rotor tending to put the machine in an attitude to decrease this speed.

The above longitudinal stability of the rotor is due to the motions of the rotor blades other than purely rotational. The rotor first of all has a coning angle, due to the fact that at all times during flight each blade is tending to adopt a position in which the centrifugal force due to rotation and the lift on the blade is in equilibrium. This position of equilibrium of the blades is a few degrees above the horizontal. In vertical descent, this coning angle is the same in all directions from the rotor hub. With forward speed, the blade on the front of the disc rises slightly, and on the rear drops below the position of coning in vertical descent. As the thrust-line of the rotor is the axis of the very flat cone formed by the blades, it is easily seen why it tilts forward below the rotor, making the rotor stable in all directions. Together with the coning and flapping, there is an oscillation of the blades about the vertical hinges. However, in ordinary flight, this is never more than two or three degrees, and has no effect on stability.

It has been shown that the rotor is essentially stable. It will now be shown that there is a definite arrangement of the component parts of the machine as a whole which gives a very good longitudinal stability.

Figure V is a side elevation of one of the latest American designs of Autogiro. This machine as a whole has many good characteristics of longitudinal stability, comparing very well with two mail planes of close to the same gross weight. As shown in this figure, the thrust-line is tilted and passes through the centre of gravity. This is to eliminate any change of balance and consequently of stability with changes in engine power.

The centre of gravity is ahead of both the rotor thrust-line and the centre of pressure of the fixed wing. In vertical descent, the nose of the machine must be held up by use of the stabilizer and elevators, otherwise the location of the centre of gravity with respect to the rotor thrust-line being as it is, the machine would take up a steep glide at a low forward speed.

In vertical descent, the fixed wing takes practically no lift. As forward speed is given the

THE PRESIDENT'S CONGRATULATIONS

President Hoover presents the Collier Trophy to Harold F. Pitcairn and his associates, April 22, 1931.

AN AUTOGIRO IN DETROIT

Autogiro of the Detroit News cruises around its home city.

machine during descent, the fixed wing begins taking a small proportion of lift, and the rotor thrust-line begins moving forward. The type of airfoil used in the fixed wing is such that the centre of pressure remains at the same position in the chord of the wing, no matter at what angle the wing is working in the relative wind. Consequently, it is apparent from the figure that the fixed wing takes a greater proportion of lift as forward speed increases and that this lifting force acts at the point indicated for the centre of pressure of the fixed wing. This counteracts the action of the rotor thrust-line, which as it moves forward tends to raise the nose of the ship.

In the machine used as illustration in Figure V, all proportions are so chosen that the tilt of the rotor thrust-line with increase in forward speed is almost exactly balanced by the nose-heavy moment due to the fixed wing. As a result, this machine can be trimmed by means of an adjustable stabilizer so that it will balance at practically any speed, "power-on" or "power-off."

Until more recent Autogiros, it was found that the high centre of gravity with the low fin area formed by the fuselage and vertical surfaces rendered the machine laterally unstable at high speed. This has been corrected in the present machines by use of substantial dihedral in the fixed wings, and turned-up wing tips, giving the effect of high fin area. Due to the proper design of fixed wing, the present Autogiros are laterally stable and follow the rudder nicely in any manœuvre (see Figures V and VI).

When the rudder is used to turn the machine, a flat turn would result with attendant skidding if no upturned tips were used. At the first tendency to side skid, due to centrifugal force F_c (see Figure VI), the pressure of the air causes a couple ($F_1 + F_2$), which swings the machine into a bank of such an angle that the righting couple caused by the weight, W, being displaced laterally at distance d, just equals the couple, F_{ck}. When the correct banking attitude is reached, the moment ($F_1 + F_2$), m, is eliminated since there is no further tendency to skid. The ship is in lateral equilibrium under the lift force, L_1, the centrifugal force F_c and the weight W. The turn is completed by the use of the rudder. Then the machine

HAROLD F. PITCAIRN

President, Autogiro Company of America

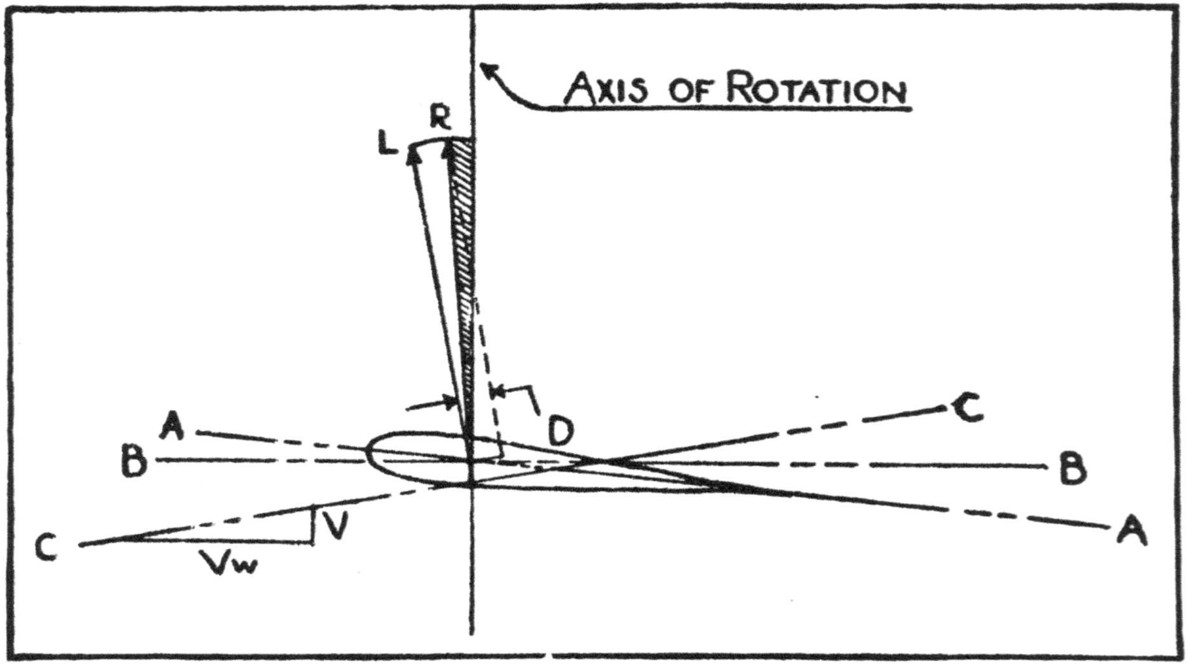

FIGURE I.

straightens its course, and the centrifugal force F_o is eliminated. The righting moment, W_d, returns the craft to a horizontal position, where $D = O$.

The amount of banking for any turn must increase with the speed of the machine and the sharpness of the turn. The correct moment for each bank is obtained automatically because a sharper turn at higher speed will result in a larger force on the upturned tip, with a steeper resulting bank, since W remains constant.

Most interesting from a theoretical point of view and most important in practical experience is the fact that the hinged system of rotating blades follows the fuselage in its manœuvres with less resistance than fixed wings follow the fuselage of an airplane. The manœuvreability of the Autogiro, as a result, is excellent, and on account of its extremely slow possible forward speed, it is capable of much shorter radius of turn than is an airplane.

The theory of the Autogiro is based on rational assumptions, which, while they have differed with the ideas of certain aerodynamists, have nevertheless proven themselves to be very close to fact in the light of full-scale performance of Autogiros.

Autogiro performance may be divided into two parts: that of low incidence, high-speed flight; and that of high incidence, low-speed flight, and vertical descent. All items for the high-speed condition of flight are referred to the fundamental parameter of ratio of peripheral speed at the tip of the rotating blade to advancing speed of the machine as a whole.

There is actually a small gap between the high-speed and low-speed portions of the calculations. However, the graphical results for the two portions of the calculations from which rotor lift and thrust coefficients are obtained may be connected by a smooth curve, with no change of slope or radius of curvature between the two conditions being connected. This, it is believed, is a good check on the validity of the theory from a purely academic point of view (see Figure VII).

In the calculations of the theory, detail derivations of various rotor features are arrived at. The power losses of the rotor or drag are shown to be the sum of frictional or profile loss, and induced drag or power loss. By the fundamental equations for these drags of a certain symmetrical airfoil, much used in rotating blades, it is possible to correct the values calculated for any airfoil of known characteristics.

The lift coefficients, downwash, and incidence can also be found in function of ratio of peripheral to advancing speed for any type of rotating system of blades.

A set of relations has been compiled by which it is possible to predict the rotational speed of a new system of rotating blades with uncanny accuracy. Knowing rotational speed, it is easy to obtain the forward speed corresponding to any value of ratio of peripheral to advancing speed. With these values, characteristics of the rotor may be tabulated in the ways that airfoil characteristics are shown.

The fact that an Autogiro is supported by a rotor rather than a system of fixed wings does not impose any abnormal loads on the fuselages or control surfaces. In fact, any loads in flight are actually less than in an airplane, as the rotor will yield under an accelerated load and dissipate a large portion of it, whereas the fixed wing machine must take all the load. Consequently the stresses on fuselage and control surfaces are less in an Autogiro than in an airplane.

The theory does not impose any limitations as to size of Autogiros. These machines are now made in sizes from 950 pounds weight empty to 3,000 pounds gross weight, with the very good speed range of six to one. While the Autogiro can land with practically no forward speed, this speed range of six to one referred to is the range of speeds that the Autogiro can maintain with no loss of altitude. It is believed by those familiar with the engineering aspects of the Autogiro that it is equally possible to build small, high-powered, high-speed Autogiros, or large transport-size machines with good high speed, with no appreciable sacrifice in low-speed characteristics.

Once a flying machine having the characteristics of the Autogiro has been achieved and is demonstrated on a practical basis, it is necessary that its performance be brought up to that reached by other and older means of mechanical flight. In this light, the situation faced by the Autogiro is most promising.

At the present time, Autogiros are from ten to

fifteen miles per hour slower in point of top speed than are corresponding airplanes. In 1925, when the Autogiro was first being flown to any extent experimentally, the highest speeds achieved were around seventy miles per hour with the expenditure of 200 horsepower. At the present time, 120 miles per hour level flight can be reached with the same power. Therefore, it can be seen that the high-speed differential as regards the airplane and the Autogiro is rapidly being diminished as the Autogiro progresses. The Autogiro is particularly susceptible to improvement, and having many more parameters from which to work than is the case with other aircraft, offers a particularly fertile and attractive field to the talents of the engineer.

XXI

A PILOT'S MANUAL FOR THE AUTOGIRO

By James G. Ray

ONE of the first interests of the professionally minded spectator of an Autogiro flight is likely to be the question as to how its operation compares with flying an airplane. On this point there has been a good deal of misunderstanding and misinformation. It has been supposed that some extraordinary skill or technique is required to fly an Autogiro. This is by no means the case. It is my opinion that a totally untrained novice could learn to fly the Autogiro in one-third to one-fourth the time needed for airplane training. It is a matter of actual experience that a competent pilot can fly it after a few minutes of instruction on the ground, and that half an hour or so in the air will make him proficient with it.

I know of no airplane pilots who cannot fly an Autogiro safely after a few minutes of instruction and suggestion before taking off, or after a little rudimentary duel in the air. Practically the only instruction required for first-class pilots is that in handling the controls for the mechanical rotor starter. The rest is simply a matter of easy practice, which quickly leads to sufficient skill for all the characteristic manœuvres of the Autogiro.

In introducing a pilot to the Autogiro, we show him first how to operate the mechanical starter. The rotor brake is first to be released; then the starter clutch is engaged. The engine should be idled and the clutch engaged slowly until the rotor starts to turn. The pilot watches the engine and rotor tachometers, keeping the relative motor speed slightly in the lead of relative rotor speed. In other words, the pilot builds up his rotor speed at a smooth pace, without attempting to force his engine.

When the tachometer shows that the rotor is making 100 revolutions per minute or more, the machine is ready for the take-off. The clutch is released and the engine throttle opened wide immediately, before the rotor has an opportunity to slow down. As the Autogiro takes off the rotor automatically accelerates to 125 or more revolutions per minute and holds that speed until the wheels again touch the ground.

So much instruction, plus some advice on placing the Autogiro properly in relation to any wind that is blowing, is actually enough to permit a pilot to take off, fly around freely and make good landings. He will experience no difficulty once in the air, if he will handle the Autogiro exactly as he would an airplane. He will find the same response to the controls while in flight, and if he comes in to a landing at a glide angle suitable to an airplane he will discover no difference except that his craft will come much more quickly to the end of its landing roll. He should make such a flight and landing before attempting vertical descent, since it will convince him that he can handle an Autogiro with as much confidence as an airplane.

The pilot should understand and discover by a

test flight that within the range of speed proper to the airplane, the Autogiro flies exactly like an airplane. The controls feel the same; the stability is the same. Whatever additional technique is needed applies almost entirely to flying in the lower speed ranges where the airplane becomes dangerous. Landing at no miles per hour is also a unique experience. At less than fifty miles an hour a pilot enjoys a new sensation, but it is one which involves no hazards. Of course, he should first experiment in the lower speed range of the Autogiro, at a sufficient height, until he is familiar with the fact that he can maintain altitude at a speed as low as 23 miles an hour.

When flying an Autogiro at speeds too slow for airplanes, the pilot will discover that his controls still operate though more sluggishly than when flying faster, and he must handle them accordingly. He will also find out that his instinctive apprehensions in regard to slow air speeds are out of place in an Autogiro. I have found that some pilots have difficulty in convincing themselves that the machine will not suddenly slip off to one side or spin. But a very little experience proves otherwise. The same applies to vertical descent, which seems to deny everything that the pilot has learned but is nevertheless a perfectly normal attitude for the Autogiro and the accepted manner of losing altitude and approaching a landing.

The best place to practise vertical descent at first is at a sufficient height from the ground so that the landing need not actually be made. Two hundred feet clearance is plenty for the purpose and all that is necessary to enter vertical descent is to idle the engine and pull the stick fully back. When the pilot has discovered for himself the novel sensation of coming down more or less vertically, he can pick up forward speed again and climb for another try at it. On account of propeller torque it is inadvisable to open the throttle in vertical descent when within ten feet of the ground, so the novice is advised, except when landing, to stay above that level until he is familiar with his craft.

It has been said that in true vertical descent the Autogiro lands itself. So it does, but sheer vertical descent is not recommended except when the pilot can take advantage of some head wind. In this case, of course, the Autogiro is landed vertically in respect to the ground but at an angle in respect to the wind. The most practical angle of descent is something less than vertical, and the actual landing should be made with a slight increase of forward speed. This assures the full efficiency of the controls, and scarcely affects the fact that the Autogiro lands and rolls to a standstill within its own length or less.

It is important to head directly into the wind while landing. Drifting cannot be kicked out with the rudder just before the wheels touch, since the Autogiro has practically no forward speed when landing in its characteristic fashion. But since it needs so little space in which to make a perfect landing, there is never any need to come in except against the wind.

If the landing is made not entirely vertically but at a slight forward angle and with a small amount of speed, the controls will remain effective. The tail is depressed in order to slow the craft as it comes down. The three-point landing is desirable though not necessary, and an Autogiro can also be landed wheels first. In any event it will come almost immediately to a stop.

When the flight is over, the rotor should be stopped. A brake is provided for the purpose. When stopping the rotor in a strong wind, the machine should be turned so that the wind comes from the left, since the rotor is tilted in that direction. This will avoid the possibility of the wind raising the blades too high as they lose centrifugal force.

The rotor should also be brought to a standstill before taxiing the craft up to the crowd. The rotor blades at the left and rear of their circle swing about waist high from the ground, and visitors to a flying field cannot be trusted to stay safely out of their way. The machine should always be placed so that the wind comes from the left or rear. If headed into the wind, a sudden gust may lift the blade too high and cause damage. By the observance of proper technique, which is more a matter of information than practice, the Autogiro can be handled on the ground in high winds without the danger of breaking blades.

There is nothing more to be said for the guidance of a good airplane pilot as he attempts his first flight in an Autogiro. In ordinary weather much less might be told him and he would be perfectly able to take off, fly around and manage all ordinary manœuvres and make good landings. By experience and practice he will rapidly improve his technique, but without such proficiency he can fly the Autogiro in entire security to himself and his craft.

One of the first pilots outside the Autogiro organization to fly the machine was Richard H. Depew, Jr. His impressions were given in a letter written shortly afterwards:

. . . "It may be of interest to you to have the impressions of one who is a complete novice with the Autogiro. Before allowing me to fly it alone, Mr. Ray took me up for about ten minutes. I had the controls for a couple of minutes in the air, doing several turns to the right and left, but not touching them during the take-off, climb or descent and landing. Mr. Ray then explained to me the method of starting the rotor and bringing it up to speed, after which I was able to taxi out, bring the rotor up to speed, take off, fly for about twenty minutes, and come in and land without the slightest difficulty.

It is my opinion that the Autogiro can be flown

by any reasonably good airplane pilot without any difficulty whatever. Furthermore, I believe that a novice could learn to fly the Autogiro more easily than an airplane. The take-off and climb are made in just the same manner as an airplane, except that it is possible to take off with a much shorter run and to climb at a very much higher angle and at a very much lower speed of advance. In the air, if I had not looked up at the rotor, I would have thought I was flying an airplane. . . . The ease of controlling and the responsiveness of the controls at exceedingly low air speed are astonishing to one accustomed to flying an airplane."

This is representative of the opinion of every pilot who has flown an Autogiro. Such critical comments as have been made are usually referable to the pilot's previous experience, according to the type of airplane he has been flying. It is natural, for example, that a man accustomed to flying small, fast ships of the pursuit type might find a large Autogiro less sensitive to the controls, for the same reasons that he would find a large-load-carrying transport airplane much different from his accustomed type of ships. The same would

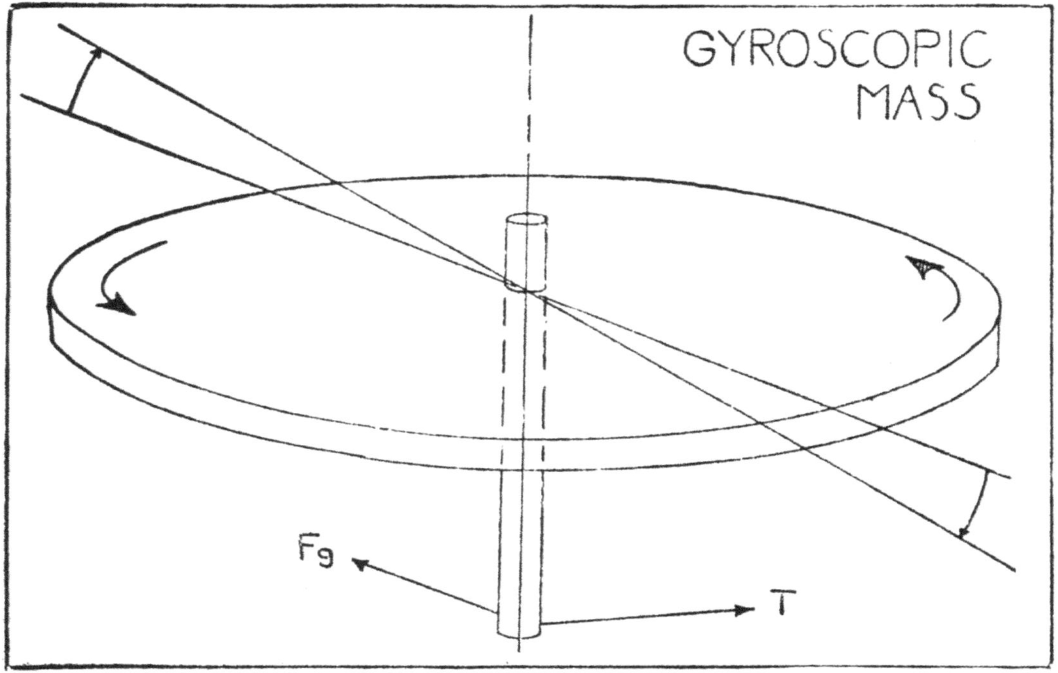

Figure II.

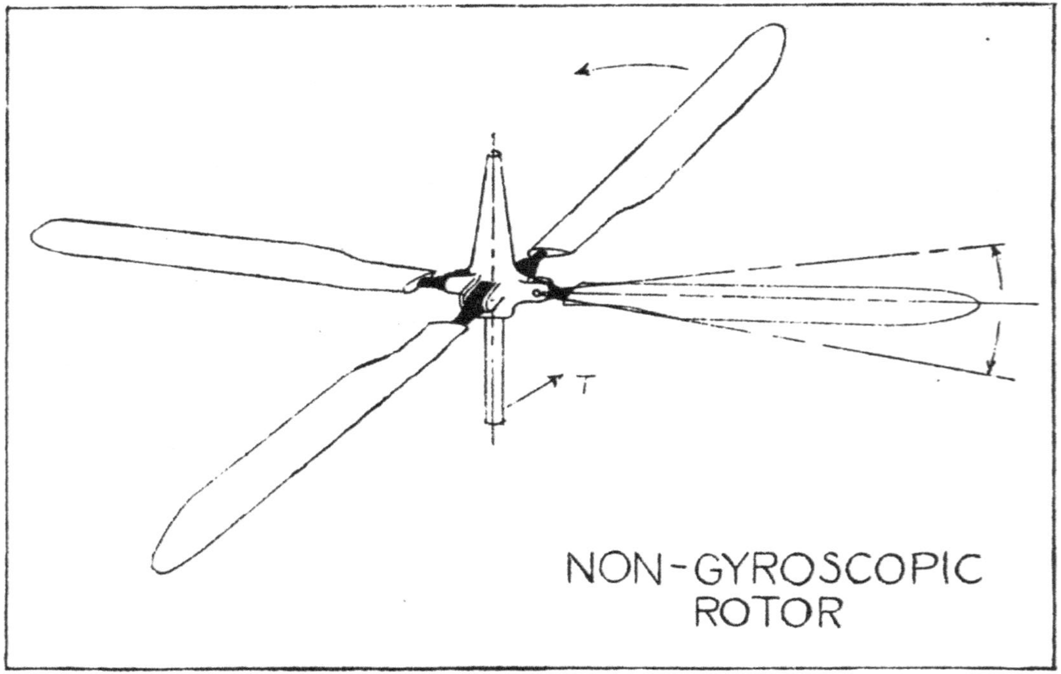

Figure III.

hold true of a big-ship pilot suddenly transferred to a small fast Autogiro, but the difficulty of transition is no greater than that encountered in changing from one type of airplane to a radically different type.

It might be asked whether increasing familiarity with the Autogiro is likely to spoil a pilot's airplane technique. I have not found it so. Even when thoroughly accustomed to slow flight, vertical descent and the other performance features of the Autogiro, the pilot does not forget the minimum safety requirements of the airplane. It is obvious, however, that a student trained entirely on the Autogiro would have much to learn before he could become competent to fly an airplane.

The advent of the new craft, indeed, has changed the whole outlook for the student pilot. After many hundreds of hours spent in student instruction in airplanes, I have found that there are two things on which instructors spend about ninety per cent of the dual period. One is avoidance of slipping into a spin from loss of flying speed; the other is the time-consuming operation of landing practice.

It is a laborious process to teach a student to execute a turn with slipping or skidding. The novice must acquire the faculty of telling in what direction or to what side the machine is slipping from wind currents on his face; he must keep the nose of the machine low enough to avoid the stall. After mastering these rudiments, he must practise for hours in the air before he can make a turn which could conceivably be called safe if executed near the ground with or without his motor.

When it comes to landing, he must be taught when to start his glide, allowing for wind and drift; he must acquire the skill needed for a fair glide, the steepness of which must be adjusted to momentary changes in the wind and allowances made for mistakes of judgment in distance. It must be drilled into the student that the moment to land is not when he wants to, but when the machine shows that it wants to by losing a good deal of its sharpness of control, and he must be shown that the choice whether to land or not to land must be made before it is too late to gun the ship over obstructions.

Then he must be taught how to lose altitude

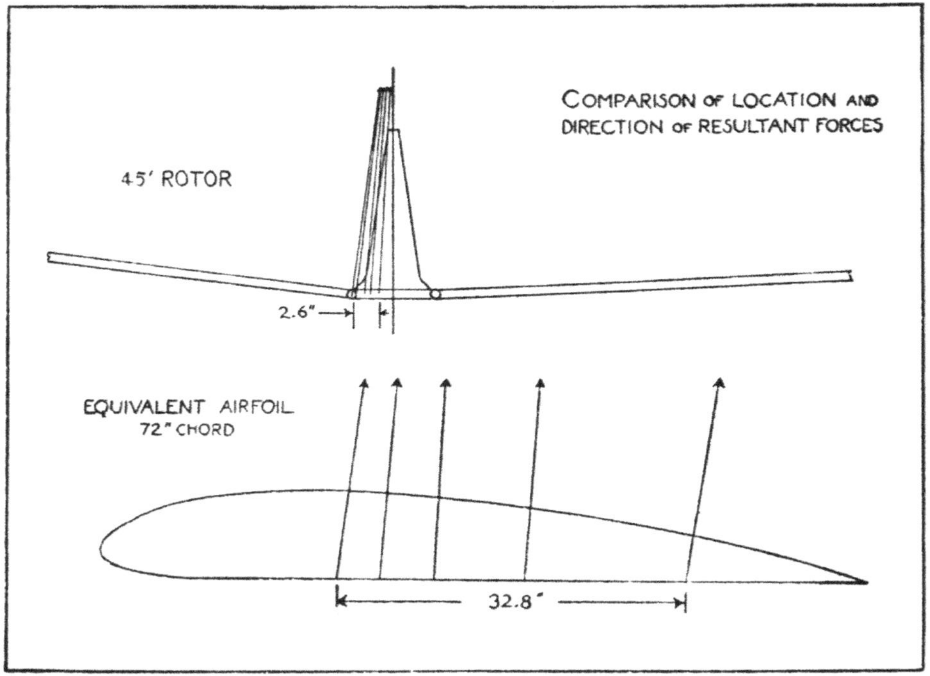
COMPARISON OF LOCATION AND
DIRECTION OF RESULTANT FORCES
45′ ROTOR
2.6″
EQUIVALENT AIRFOIL
72″ CHORD
32.8″

Figure IV.

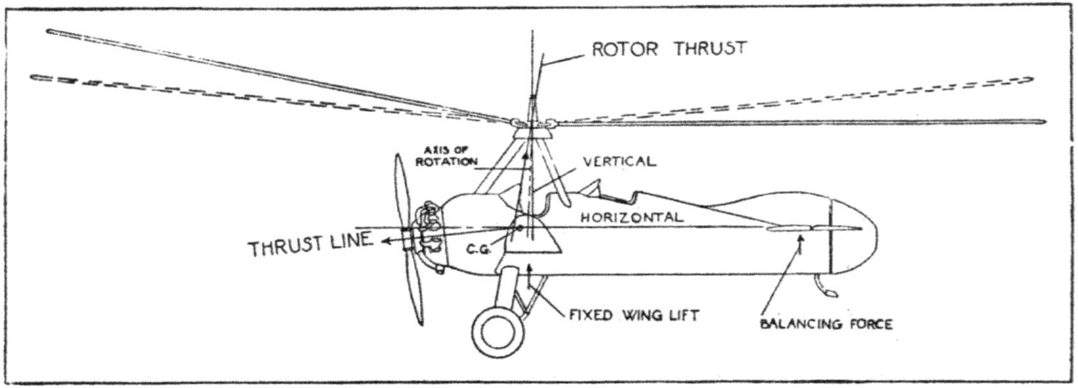
ROTOR THRUST
AXIS OF ROTATION
VERTICAL
HORIZONTAL
THRUST LINE
C.G.
FIXED WING LIFT
BALANCING FORCE

Figure V.

safely. After the first solo period, instructors spend a lot of time in imparting the skill required for S-turns, spirals and slips to a spot landing. Overshooting and undershooting, particularly in practice for landing on small and unknown fields, must be eliminated by hours of careful coaching.

All this is reduced to a minimum or made altogether unnecessary in instructing a novice in the Autogiro. Nearly ninety per cent of the problem of student instruction vanishes. The student is safeguarded from his own mistakes by the inherent nature of the machine. An Autogiro cannot spin, it practically lands itself from any altitude, it can be put down nearly anywhere, it demands no skilful estimate of speed or gliding angle, and in practically any emergency condition it can take care of itself and of the man who flies it.

For these reasons the instruction period may be shortened to as little as one-fourth of the time needed to prepare a student for solo flight in an airplane. Subsequent practice can be carried on without the supervision of an instructor. Once the rudiments of flight are mastered, the man who flies an Autogiro can perfect his own technique and performance at his own discretion and convenience. And in the meantime he can fly safely for business or pleasure whenever air conditions are reasonably favorable.

For a while there may be Autogiro schools for those wishing to learn the management of the new flying machine. But soon, I think, they will be considered unnecessary. What instruction is needed will be given as a form of service by the manufacturer to the purchaser, or else the young pilot will get his flight training from a friend or some associate on a flying field. In the early days of the automobile, there were many schools to teach the owner to drive. But very few learn how to drive today in an automobile school. They learn the essentials from somebody who knows, and acquire proficiency by actual practice in running a car, first over quiet roads under good conditions and gradually attempting the difficulties of traffic and bad weather. Something of the sort seems the logical future of flight training in the Autogiro.

This refers chiefly, of course, to the private owner and operator. For the professional Autogiro pilot there will continue to be a need for schooling in the several sciences related to flying —meteorology, navigation, engineering, etc. But the flight training that is at present so expensive and prolonged will be greatly reduced, lightening the heavy burden of expense at present borne by the student pilot. This, I believe, is profoundly important to the future of aviation. For the flying machine can never become a universal utility while it takes so long to become proficient in its use and while security depends so much on superlative skill and long experience in the man who operates it.

The Autogiro, in a word, is easy to fly. Any good pilot can learn to handle it in a few minutes; any intelligent student can master its fundamentals in two or three hours. This is not the least striking of its many advantages over the airplane which it is destined to supersede for so many of the functions of flight.

Note: James G. Ray is a pilot whose experience covers practically every phase of modern aviation. Following wartime service in the U. S. Air Corps, he was for some years successfully engaged in barnstorming, instructing, and all-round service on various types of machines for Curtiss Flying Service. When Pitcairn Aircraft, Inc., was formed, he served as chief test pilot. In May, 1928, when Pitcairn Aviation, Inc., began the operation of the New York-Atlantic air mail route, Mr. Ray became Operations Manager. Five flying schools were opened under his supervision. Since 1929, Mr. Ray has served as chief pilot in connection with Autogiro development, and is a vice-president of the Autogiro Company of America.

XXII

TRIBUTE AND TESTIMONY

THE advent of the Autogiro aroused immediate and widespread interest among American authorities on aviation and in both the technical and popular magazines. The various aeronautical journals discussed it at length and in detail, and had done so while it was still in the early stages of development in Spain and England. Possibly the first technical discussion published in America appeared over the name of my friend Heraclio Alfaro, in *Aviation* for April 9, 1923. Since then nearly every aviation publication in the United States and many scientific and engineering journals have printed articles concerning it.

It is believed that the first article in an American

magazine of general character was "The Windmill that Flies," by D. Rose, in the *North American Review* for June, 1929. This was several times reprinted, once in a German version. Other articles for non-technical readers have appeared in the *Saturday Evening Post, The Forum, Fortune*, the *Readers' Digest*, all the aviation trade journals, and several of the magazine supplements to the American Sunday newspapers.

Since the Autogiro was presented in public demonstration as a craft perfected to the point of commercial usefulness, a great variety of comment concerning it has been published. American leaders in aviation, experts in other fields, and journalistic authorities are briefly quoted in order to show the diversity of favourable opinion it aroused and the varied possibilities that presented themselves to those who studied its earliest demonstrations in the United States. The opinions of the first pilots to fly it in America are also worthy of brief record.

* * * * *

Captain Frank M. Hawks, famous cross-country pilot who has established numerous records for point-to-point speed flying in the United States, reveals in the *U. S. Air Services Magazine* his impressions of flight in the Autogiro aircraft:

"The Autogiro in its present state has sufficient perfected detail to make it absolutely safe for the average aviator to fly. It offers the smoothest, safest landing of any aircraft I have ever flown.

"The theory of the plane is absolutely revolutionary and I am somewhat enthusiastic about its development because of this fact. In aviation we need more individuality in design and the Autogiro certainly is a very marked example of this.

"The take-off in itself is effected exactly like that of an airplane. One may dive and climb at will, make vertical turns, and so on. The Autogiro flies exactly like any ordinary airplane. The real sensation that I experienced was in landing. Cutting off the motor 1,000 feet directly over the field, I pulled the stick back and proceeded to fly, on my first landing, about thirty miles an hour coming in to land. The descent was at about an angle of 45 degrees and on touching the ground I did not roll more than ten feet without any application of the wheel brakes whatsoever. I took off again and again and landed to become accustomed to this new strange craft. It intrigued and thrilled me.

"I now had enough experience to attempt a more vertical descent. This I did. There is no question but that it is quite a sensation for a pilot who has been flying a fast airplane that lands between 60 and 70 miles an hour and that must be manœuvred carefully into a field, making sure his judgment of approach is accurately managed, to step into a machine and fly right over the centre of the field, shut off his motor and then drop right straight down in the circle which marks the centre of airports.

"I believe the Autogiro has made marvellous progress and the closing chapter of my experimentation was certainly proof as to this. We flew in formation, I with the Texaco 13 (Captain Hawks' famous racing monoplane), and two pilots each with an Autogiro, from Pitcairn Field to Trenton, N. J. This proved to me that they did move along at a fair rate of speed in their flights across country, and there is absolutely no question concerning their ability to land in a restricted area.

"I want to fly the Autogiro more. The technical

side of its construction and its theory are most interesting and offers a great possibility to aviation advancement. Something that will descend vertically and rise almost vertically has long been desired.

"To land in vertical descent! Think of it! And not roll a foot forward. That is something more than airplanes can do and is worthy of a lot of consideration."

Amelia Earhart, one of America's first ladies of the air and the first to fly the Autogiro, following a series of exhaustive test flights at Pitcairn Field, Willow Grove, Pa., said:

"The Autogiro has wonderful characteristics of safety and ease of control. These characteristics should be of real influence in bringing women into aviation, and that is one thing aviation needs.

"The automatic stability of the machine, as well as its peculiar properties of safe vertical descent, make it of immense utility in all types of commercial and sport flying. It should prove an effective medium of increasing the everyday applications of air transportation, and provide a powerful impetus for business in the aviation industry.

"I found the machine easy to fly, and moreover I enjoyed flying it. Modern Autogiros have the smooth touch and sureness of control which experienced pilots look for in all well engineered aircraft."

The Hon. Hiram Bingham, U. S. Senator from Connecticut, speaking by radio over the Columbia Broadcasting System, said:

"The Autogiro is the greatest development in aviation since Orville Wright taught men to fly. The performance of landing on the little lawn in the White House grounds and taking off again from the same small space and over the trees, shows the tremendous possibility of development which this type of flying machine possesses. Never before have we been able to see clearly in the distance the business man coming from his home, taking to the air on a three- or four-acre lot. But that time now is soon to come, and the development of the Autogiro is bringing it nearer.

"President Hoover is tremendously interested in this development, because as Secretary of Commerce, he did more than any other man to develop commercial flying in this country and it was under his administration as Secretary of Commerce that the Commercial Flying Act was passed.

"The Autogiro, involving an entirely new principle, is said to have influenced some of the birds watching it to wonder if they had the right idea. I have seen a picture of two sparrows on a roof, discussing the matter and stating that they thought they must have missed the real secret. . . .

"The Autogiro offers tremendous possibilities without large airports. That is why business men can commute from within 90 to 100 miles from the centre of the city and land on the centre of their roof or in any part of town, and can take off from the roof without being in any danger of sideslipping or injuring any one. No lives have ever been lost with this and no serious crashes have taken place. We are really on the threshold of a new era."

Assistant Secretary of the Navy for Aeronautics, the Hon. David S. Ingalls, delivered an instructive and entertaining radio address on "the latest aviation development, the Autogiro," to the Ohio School of the Air, on March 16, 1931.

Secretary Ingalls briefly surveyed the history of man's attempts to fly, the humorous, the tragic phases of it, and finally the accomplished fact when the Wright Brothers flew at Kitty Hawk. Then came the Autogiro of Cierva, "the first one to develop an original means of flying since flying was actually established."

Continuing from the Secretary's address: "The main thing claimed for the Autogiro is that it can land more safely than any other sort of flying machine. . . . The alpha and omega of human conduct is safety. And so it is with aviation. The airplane is not uncomfortable or distasteful in and of itself as a vehicle of transportation. The airplane simply fails to guarantee the individual that chance of continued existence that one demands. And until travel in the air is as comparably safe as by train, ship, or automobile, travel in the air will be spasmodic.

"The Autogiro seems to be the missing link. By it one is transported in the air, and by it one is

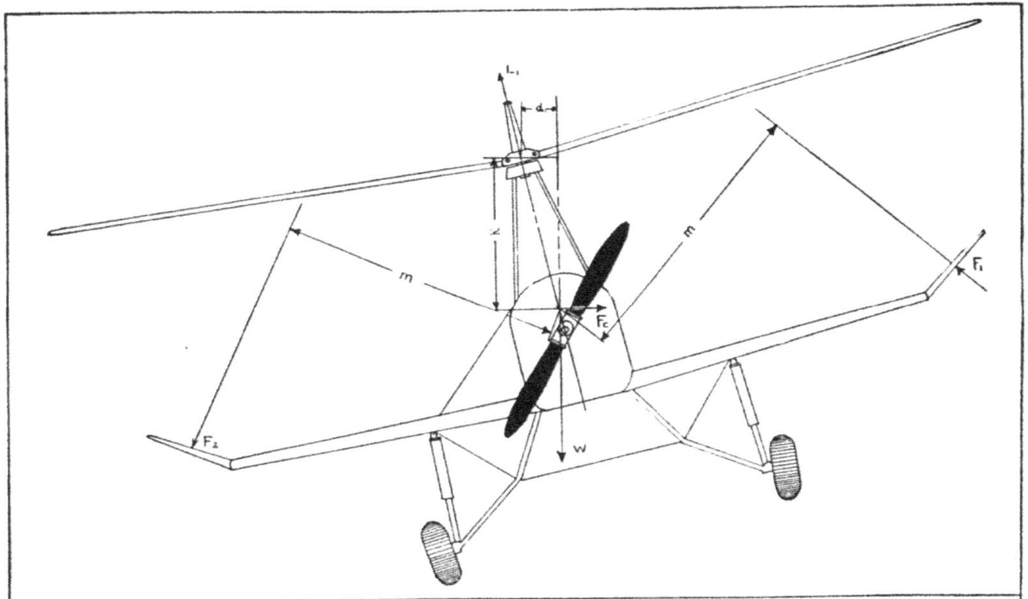

FRONT VIEW MAKING A TURN TO PILOTS RIGHT

FIGURE VI.

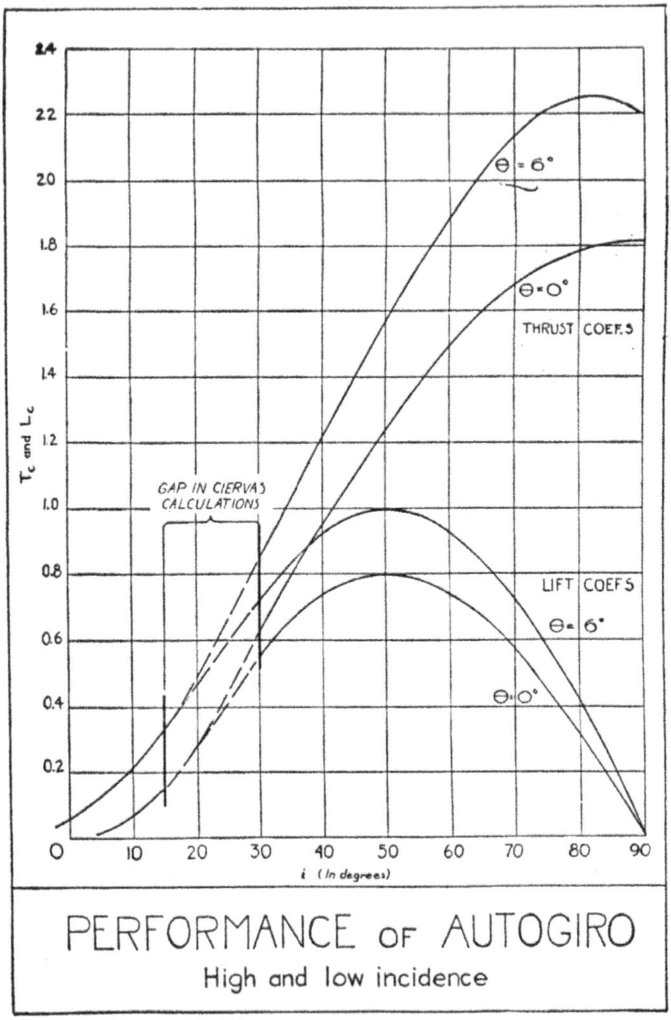

2.4
2.2
2.0
1.8
1.6
1.4
1.2
1.0
0.8
0.6
0.4
0.2
0
10
20
30
40
50
60
70
80
90
T_c and L_c
i (In degrees)
GAP IN CIERVAS CALCULATIONS
Θ = 6°
Θ = 0°
THRUST COEF.S
LIFT COEF.S
Θ = 6°
Θ = 0°
PERFORMANCE OF AUTOGIRO
High and low incidence

FIGURE VII.

transported as safely as by other means of travel. . . . Regardless of any conceivable disadvantages, this much is certain, that the pilot of this flying machine should be able to land in emergency almost anywhere with safety.

". . . In transportation on the ground man has accomplished the great speed of today not by means of a machine with movable legs, but by means of wheels. He did not follow any animal in this, but departed from the example that nature afforded. And the same result seems to be occurring in aviation. The first departure was from a flapping wing to the stiff, immovable wing of an airplane. The second from that to a wheel, for after all the Autogiro is suspended by a wheel.

"The Autogiro can be flown more easily than the airplane. And as one can in a pinch simply sit still and let it descend vertically without much danger of personal damage, it may be more readily flown by private owners than is the airplane. A very few hours of instruction and one may fly this ship home. As it can take off in at the most 100 feet, and land in half that, the Autogiro can be operated from a field only a fraction as large as is necessary for the airplane. It is literally a backyard flying machine."

Newton D. Baker, III, son of the former Secretary of War for the United States, says:

"I have, in the past five years, flown in almost every type of airplane, in half the countries of the world and although comparatively inexperienced know something of the why and how of flying.

"Recently in Philadelphia, I watched a demonstration of the Autogiro and had a flight in it. Not being a technical man, I leave that kind of description to those who are better fitted, but I have been begging for this opportunity to say that in a lifetime of interest in aircraft, I for the first time have seen a machine which inspired such confidence that I believe the novice like myself can hope to fly in safety. Its ability to take off slowly, fly slowly, and land vertically, thus overcoming the primary dangers of airplanes, certainly gives it the place of the first fundamental development in flying since the Wright Brothers. I have great hope that this amazing new machine will convince the public, as it has me, that the day of the private owner and pilot is here. Now we can fly without worry and go where we want to go. I'm saving my hard-earned pennies!"

In an article on "How the Autogiro Flies," in the *Scientific American,* October, 1929, Earl D. Osborn says:

"The Autogiro is essentially easier to fly than an airplane. It cannot be stalled or spun. There is no difficulty about overshooting a field. The angle of descent is so steep that this requires no delicate judgment nor estimating of distance. If there is a crash, it will be at such slow speeds that little damage is likely to occur. Due to its ability to fly slowly without danger, the Autogiro can be flown in much thicker weather than an airplane.

"To those who have become accustomed to flying airplanes, their first flight in an Autogiro is apt to be a little startling. When the Autogiro is throttled back and the machine apparently stops in the air, it is hard to realize that this is a normal and safe procedure and not a stunt requiring great skill. We have become so accustomed to the necessity of landing fast and of judging height accurately that it is hard to visualize how much easier it would be to fly and how many more people would undertake it if these two elements could be eliminated. The Autogiro has set new standards of performance which are bound to stimulate flying."

E. R. Fenimore Johnson, prominent explorer and former executive vice-president of the Victor Talking Machine Company, contemplating an expedition to unexplored regions of South America to make the first genuine sound pictures of jungle life, recently was given a demonstration of the Autogiro.

Mr. Johnson, who is himself an amateur pilot, was enthusiastic over the new type aircraft's safety features and advantages for private owners as well as its possibilities for exploring expeditions.

"The Autogiro would vastly increase the range of effectiveness of an expedition. Its ability to come down low and fly at slow speeds would give an observer wonderful opportunity to study the nature of the country, its geological outcroppings and possible archaeological remains. Also it would be unexcelled for taking photographs and spotting human and animal life in the dense jungles.

"But perhaps one of the most important uses of the Autogiro would be in such emergencies as accident or sickness among the men. The craft's ability to land straight down and take off from restricted areas would make it easily possible to keep in touch constantly with field parties. Moreover, with the machine's ability to hover, mapping would be greatly facilitated."

An entertaining but complimentary comment from the *New Yorker,* November 1, 1930:

"In the interests of aviation we allowed ourself to be wafted gently up over New Jersey last week, in an Autogiro. We felt like a dandelion seed. We even took hold of the Autogiro's controls, and felt even more like a dandelion seed. This rare device of the upper regions is not airplane, not balloon, not dirigible, not helicopter—it is a Spanish windmill plane, that holds you up largely by a four-bladed horizontal fan, slowly rotating.

It is like an airplane that has been soothed and calmed down; it is like an airplane to which someone has said: 'There, there.' "

"You can be a pretty slipshod pilot and get away with it. Suppose you take off to fly over a barn, and find you aren't going to make it. No problem at all—you pull the stick back and the contraption shies away from the barn the way a jumping horse shies away from a high fence after running at it full tilt.

"We hear that the Autogiro is to be manufactured commercially here soon, and will sell for about what a plane of like power sells for. And although it looks like something Jules Verne thought of, it will actually land in one's flower garden—or, if one is fussy, in one's neighbor's flower garden. We were much impressed by our flight, and we want an Autogiro more than we have wanted anything since we wanted a pair of rabbits."

"The Autogiro almost satisfies the exacting demand of a leader in the aviation industry for an airplane which can take off from the roof of a skyscraper and land on Pike's Peak. If not entirely foolproof, it is nearly so. Practically any one, with sound facilities, can fly an Autogiro, although an Autogiro pilot might not have the skill to fly an ordinary airplane.

"It makes aviation almost a backyard proposition, because it can land easily in a lot seventy-five feet square, and take off in not much more space. It could land, for instance, with ease in Fifth Avenue in front of the Public Library and take off from the same place.

"In short, it is what every one is looking for, the practically safe airplane."

Bruce Gould in the *New York Evening Post.*

Colonel Leonidas Coil, Chief Fire Warden of the State of New Jersey, who in the work of combatting destructive forest fires has spent hundreds of hours in an airplane directing his forces, has investigated the possibilities of the Autogiro for this type of work. He says:

"Fire fighting over densely wooded districts has only been practical because of the wonderful reliability of modern airplanes and motors. With the Autogiro aircraft, which in the event of forced landing from motor trouble can settle vertically into small clearings, the danger of forced landing would be brought down to a minimum. Then, too, the man in the machine could maintain better contact with his ground force than is now the case, when the only communication possible consists of ground signals and messages dropped from the ship. The Autogiro should have immense possibilities in this as well as in other types of flying, where security is the most important consideration.

"Speed has always been one of the chief considerations in flying, and in this respect Autogiros approximate the performance of other modern aircraft. However, the ability to fly at slow speed also has advantages beside the more obvious one of safety. It makes possible for one thing the careful mapping and inspection of terrain which are of immense benefit in such work as fire fighting."

The Buhl Aircraft Company, with offices in the Buhl Building, Detroit, and plant at Marysville, Mich., is a pioneer organization of the aircraft industry which has long been known for its conservative business attitude and the record-breaking exploits of the airplanes it manufactures.

"A nation-wide clamour for the Autogiro" as shown by a questionnaire sent recently to 50,000 persons, crystallized their decision to manufacture aircraft of the Autogiro type, and the Buhl Aircraft Company became in March, 1931, the third to enter the Autogiro field.

In announcing their decision, the Company made the statement:

"There is no question in our minds but that the Autogiro answers one of the great problems of the aviation industry—getting the older generation into the air."

From the magazines and newspapers:

(Extract from Edison's answers to *Review of Reviews* questionnaire)

Q. Do you think the Autogiro is the coming thing in aviation—the best principle so far developed?

A. Yes, and it came from Spain. They say that Spain is dead. But that man (Cierva) has the egg of Columbus.
Associated Press release, Dec. 24, 1930.

"Airplanes challenge older transportation and the Autogiro challenges the plane."

Raymond Willoughby in *Nation's Business* for March.

"Experts regard the Autogiro as the logical vehicle of the air for the vast majority of the potential flying public."

Atlanta (*Ga.*) *Journal*, Jan. 16, 1931.

"I have just had the biggest thrill of my twenty years of flying. I have piloted an Autogiro. And I have seen this amazing plane 'do the impossible.'

"It is, I am positive, the flying craft of the future."

Assen Jordanoff in *Popular Science*, March, 1931.

"Development of the Autogiro will open up new

fields of exploration which can be made only by airplane."

Commander Richard E. Byrd, in the Providence (R. I.) *Bulletin.*

"The ordinary plane has progressed to approximate perfection by steady development in details but is still the recognizable descendant of the frail craft of twenty-five years ago. The Autogiro has borrowed a wealth of experience from conventional aeronautics but is a fundamentally new type of flying machine probably only at the beginning of its real development."

The Philadelphia *Evening Ledger.*

The Autogiro was given the ranking position for modernity in aviation by consensus of opinion among aviation leaders and automotive engineers, says a news story in the New York *Sun* for March 19:

"A committee in charge of corner-stone laying ceremonies for the Hampshire House, a new thirty-seven story, $6,000,000 modern residential hotel in process of construction at 150 Central

Park South, New York, interrogated a large group of distinguished men and women in varied fields of art and science. The purpose of the committee was to secure opinions on the most representative modern productions to be placed in the cornerstone.

"In the fields of art and literature there was great divergence of opinion, but in the field of aviation the Autogiro was selected without competition. A photograph of the latest model will be placed in the stone."

"This promises to be an Autogiro year in aviation. The peculiar airplane with a windmill above it has been subjected to intense study and experimentation for the last four years and now it will be placed before the country as the safest and best method of air travel. . . .

"Its advocates claim it can be flown safely by the average man or woman after a short course of instruction. It lands slowly and easily and can be brought down almost vertically. It takes off almost as easily. It is declared stable and efficient in the air. . . .

"If it is a safe machine for the average man or woman to use, its advent will stimulate the aviation industry remarkably. There are thousands of men and women who would take up flying if convinced they could do so with maximum assurance of safety."

Syracuse (N. Y.) *Post-Standard, March* 18.

"The Navy has just bought an Autogiro—that remarkable airplane which can take to the air and land on the ground in an almost vertical line and at a leisurely rate of speed. The added safety and comfort which the Autogiro would bring to civilian flying are plain, and if it could add to its own peculiar gifts those now belonging to the fast military pursuit plane its great utility as a weapon would be uncontested. Every cruiser and battleship could be equipped with several of them and the immense landing and taking-off decks of the craft carriers would become obsolete, inasmuch as the Autogiro can drop like a feather to a given point. . . .

In the land operations which the Navy and more particularly the Marine Corps, undertakes the

Autogiro seems to have brilliant possibilities. . . . With the present style of airplane the pilot could not always pick the spot which was the safest with regard to the enemy's position on the ground. . . . On many occasions the planes could not land at all, thereby making it impossible to remove the wounded or bring reinforcements of men, although weapons and munitions could be dropped to the ground. It is therefore pertinent to note that the Navy plans to experiment with the Autogiro on land as well as on the sea. It will be surprising to the civilian if these experiments do not result in some interesting innovations in the military art."

New York *Herald Tribune*, January 24, 1931.

The Aviation Collection by Sportsman's Vintage Press

www.SportsmansVintagePress.com

Aircraft Construction Handbook	by Thomas A. Dickinson
Aircraft Sheet Metal Work	by C. A. LeMaster
The Aircraft Apprentice	by Leslie MacGregor
Aircraft Woodwork	by Col. R. H. Drake
Aircraft Welding	by Col. R. H. Drake
Aircraft Sheet Metal	by Col. R. H. Drake
Aircraft Engines	by Col. R. H. Drake
Aircraft Electrical and Hydraulic Systems, and Aircraft Instruments	by Col. R. H. Drake
Aircraft Engine Maintenance and Service	by Col. R. H. Drake
Aircraft Maintenance and Service	by Col. R. H. Drake
Wings of Tomorrow: The Story of the Autogiro	by Juan De La Cierva & Don Rose
Aeroplane Engines in Theory and Practice	by John B. Rathburn
Aircraft Radio and Electrical Equipment	by H. K. Morgan
Aircraft Engine Design	by Joseph Liston

Made in the USA
Monee, IL
05 November 2026

67816235R00187